Multi-Airfoil Design:
Spectral Approach

Namita

Table of Contents

Chapter 1

Introduction

1.1 Motivation

The inspiration behind this research stems from the growing drive from industry and academia to understand the flow physics for low-Reynolds number flows over multi-element wings. Through advancements in computing power, more advanced numerical approaches such as the high-order spectral element approach can now be implemented for more accurate flow modeling within a practical time frame. In addition, the availability of a spectral element Direct Numerical Simulation (DNS) code, Nek5000 [1], enables simulation of complex geometry flows like that over a multi-element wing with high precision rates and easy scalability to higher Reynolds numbers without unnecessarily sacrificing the time of generation of grids. Ultimately, the availability of water tunnel results [2] [3] for flow past multi-element wings also makes this research a validation effort using Nek5000 to simulate aerodynamic flows.

1.2 Multi-Element Airfoils

High-lift devices are used to boost aircraft performance while providing additional lift during take-off and landing.

Over the last few decades, the use of high-lift configurations has become one of the major parameters influencing the entire design of the aircraft, in particular the cost. The high-lift system, first, is sophisticated and costly. As per Rudolph [4], the high-lift configuration represents 6 to 11 % of the total cost of manufacturing a jet. Second, its success has a significant impact on the efficiency of the aircraft. According to Butter [5], for a conventional twin-engine jet, a 5 % rise in maximum take-off lift coefficient will increase the payload of the aircraft by 12-15%. Similar examples have been given by Garner et al. [6]: an increase of 1.5% in the maximum lift coefficient is equivalent to an increase of 6600 lb in payload at fixed approach speed, and an increase of 1% in the lift to drag ratio during take-off is equivalent to an increase in payload of 2800 lb.

All of these examples illustrate the importance behind the design of a high-lift configuration. High-lift architecture and lift-enhancing mechanisms are continuously being redesigned in order to achieve higher capability, both in terms of aerodynamics, and hence fuel consumption, and more recently, noise abatement. As per Van Dam [7], early efforts were based on optimizing the lift coefficient to meet the transport aircraft's high cruise wing loading needs while maintaining reasonable takeoff and landing distances. The complexity of the structure presents unique challenges to the aerodynamic design: high lift elements must be deployed in take-off and landing and yet be stowed cleanly for aerodynamic optimization of cruise conditions.

The high-lift system for most commercial airliners, such as the Airbus A320 and the Boeing 737, consists of two components: a slat (leading-edge element) and a flap (trailing-edge element). Figure 1.1 illustrates the increased performance characteristics that result from use of these two devices: Flap deflection (in green) increases lift coefficient (by ΔC_L) by essentially increasing overall camber of the airfoil, while slat deployment (in red) acts as an increase in angle of attack (by $\Delta\alpha$). Together, the deployment of flap and slat result in a significant boost in lift coefficient (upper dotted line in Figure 1.1).

The flow physics for high Reynolds number flows ($O(10^6)$) past high-lift con-

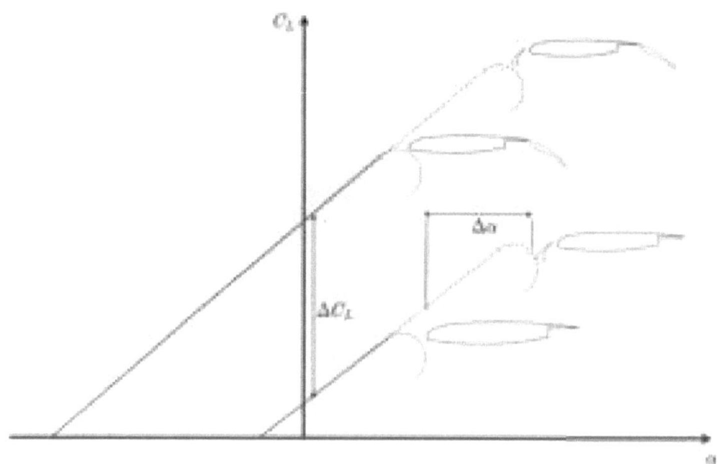

Figure 1.1: Effect of a conventional high-lift configuration on aircraft lift

figurations is dynamic and hard to predict as shown in Figure 1.2. The complex geometry can give rise to multiple separated boundary layers, separation bubbles, transition regions, and wake/boundary layer interactions [8]. While several experiments have been performed, there is still a long way to go to gain a full understanding of the inherent complexities.

1.3 MDA 30P30N IIigh Lift Airfoil

Several multi-element airfoils have been studied in the literature: among them, the McDonnell Douglas Aerospace (MDA) 30P30N three-element airfoil stands out. The 30P30N three-element airfoil as a typical configuration, shown in Figure 1.3, has been extensively analyzed both experimentally and computationally. A review of investigations on this three-element high-lift airfoil is given in the following sections.

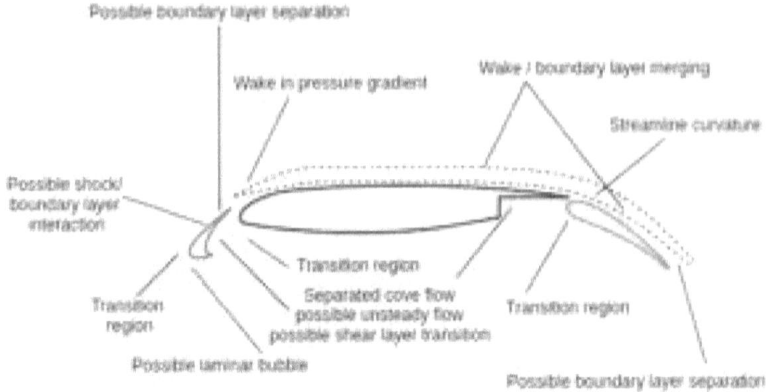

Figure 1.2: Flow physics for a conventional high-lift configuration

Figure 1.3: MDA 30P30N three-element airfoil

1.3.1 Experiments

There have been several experiments on the 30P30N three-element airfoil to gain more knowledge of high-lift flow dynamics and to provide validation data in the literature for various Computational Fluid Dynamics (CFD) codes.

The 30P30N airfoil was tested for two Reynolds numbers, 5×10^6 and 9×10^6, at the NASA Langley Low Turbulence Pressure Tunnel (LTPT) by Chin et al. [9]. The authors found an increase in the lift with a rise in the Reynolds number. This propensity increases with increased angle of attack, α, but the stall angle is approximately the same for both cases. The pressure range at $\alpha = 4°$ suggests that for the smaller Reynolds number separation on the flap is more likely to occur.

Also, the lower Reynolds number flow has a tendency to move the wake farther away from the upper surface of the flap.

The skin friction for the 30P30N airfoil was measured by Klausmeyer et al. [10] for three different Reynolds numbers: 5×10^6, 9×10^6, and 16×10^6. Test findings indicated that the stall is triggered by reversal of flow in the wake of the main element, rather than by separation on the flap surface. Strangely, flow separation was observed on the flap at a low angle of attack however no separation was observed with an increase in the angle of attack. This flow separation on the flap is dependent on Reynolds number and this finding is in harmony with the results of Chin et al. [9].

Although the 30P30N airfoil is a well-known standard 2D system and is tested as a 2D example, one main problem CFD studies have posed is the inconsistency between computational simulation and experiment. In addition, wind tunnel experiments also entail uncertainties such as effects due to supporting structures, wake caused by the tunnel wall, and 3D effects of the wing.

Bertelrud [11] and Rumsey et al. [12] analyzed these three-dimensional effects. A fully two-dimensional, high-lift experiment cannot exist according to Bertelrud [11]. The presence of side-walls reduces wing circulation in the wall-juncture area and three-dimensionality can ruin the flow, particularly on the flap, as the angle of attack increases further. Having sidewall ventilation can decrease this effect but it can never be fully removed. In fact, on the flap at angles of attack beyond $16°$, the three-dimensionality was substantial even though the side-wall vent was present.

1.3.2 Computations

Much computational research on the 30P30N airfoil has been carried out since NASA Langley first set a high-lift CFD challenge in 1993 focused on this geometry. Almost all of these studies were based on RANS codes, using structured or unstructured meshes, and conducted at a Reynolds number of 9×10^6. Two aspects were pointed out: the behaviour of the turbulence model and disparity between experimental and numerical results.

Various turbulence models have been used and compared in several computations, and the Spalart-Allmaras (SA) turbulence model is the most frequently used. While Lynch et al. [13] suggested that the SA turbulence model was not satisfactory to model flow over the 30P30N configuration, Klausmeyer et al. [14] successfully modeled the flow using the SA turbulence model with a 2D RANS code with preset boundary layer transition locations as the SA turbulence model is not suitable to predict the transition location.

Multiple turbulence models such as Spalart-Allmaras (SA), Baldwin-Barth (BB), and Shear Stress Transport (SST) models were compared for the 30P30N high-lift system by Rogers et al. [15]. Discrepancies between various turbulence model estimates were smaller than discrepancies between CFD and experimental results [8]. The flow across the flap was estimated by Valarezo et al. [16] using three separate turbulence models: BB, SA, and k-ϵ. They came to a similar conclusion mentioned by Rumsey et al. [8]. All these turbulence simulations under-predicted the position of the reattachment point.

There have been numerous studies to figure out how conventional CFD methods can estimate the variations (of drag, lift, transition location, etc.) caused by changes in the Reynolds number. In 1994, Dominik [17] also performed incompressible RANS simulations with a structured grid code using three models of turbulence: BB, SA, and SST. According to him, the BB model was able to predict significant changes in the flow resulting from a change in Reynolds number, whereas the SA and SST models could not detect any changes. However, Jones et al. [18] reported contrasting outcomes by performing simulations with the SA model using a RANS code on structured grids and revealed that the SA model is capable of estimating changes due to variation in Reynolds number.

The SST and SA turbulence models were not developed to anticipate transition, as is the case for other models in common use today. The uncertainty around using RANS modeling for this flow has been discussed in the literature. This flow presents quite large regions of laminar flow where the use of turbulence models is questionable [19]. The setting of the far-field boundary location was found to significantly influence the estimation of drag. A correction for far-field propagation was needed to increase the accuracy of RANS modeling if the far-field is positioned closer than 40 chord lengths away [20].

Bodart et al. [21] implemented a large eddy simulation (LES) model to predict flow around the 30P30N three-element airfoil. Their results showed a good agreement with experimental data for aerodynamics coefficients. In 2005, Langtry and Menter [22] applied the $\gamma - Re_\theta$ turbulence model to predict transition location on all the elements of a 30P30N high lift airfoil.

As noted from the investigations mentioned in previous sections, these high lift devices are not yet explored numerically for low Reynolds number ($O(10^4)$) regimes. Also, previous studies at high Reynolds number ($O(10^6)$) were carried out using RANS, LES, and hybrid turbulent models which have difficulty exactly reproducing experimental results, due to a lack of detail of important flow physics, namely transition. We therefore seek to develop direct numerical simulations for high lift configurations to provide more accurate representation of the complex flow physics and to study the differences between two-dimensional and three-dimensional flow for this geometry. This will help to confirm whether turbulence models, such as those mentioned above, can be used effectively, once we understand the basic flow physics that might be missed by the models. This work represents a first step towards direct numerical simulations of high lift devices at high Reynolds numbers ($O(10^6)$). We aim to study lift and drag characteristics, instability modes and coherent structures as a function of varying Reynolds number.

1.4 book Outline

This research aims to provide a detailed explanation of low-Reynolds number flows over a 30P30N three-element high lift wing by way of computational simulations using a high-order spectral element method (SEM). To address this investigation effectively, this study is split into chapters as follows.

Chapter 2 describes the previous work related to low-Reynolds number flows over the 30P30N wing, which forms the point of departure for the current research. Chapter 3 outlines the methods utilized from mesh generation to post-processing for high-order spectral simulation of flow past a 30P30N three-element wing. Results obtained and comparison with available experimental data are discussed in Chapter 4 for both 2D and 3D SEM simulations. Finally, Chapter 5 provides the conclusion of the present work and provides suggestions for future work.

Chapter 2

Background and Literature Review

For high Reynolds number flows ($Re_c = 10^6 - 10^7$), the 30P30N three-element high lift airfoil is widely studied numerically and experimentally [7] . Nonetheless, low Reynolds number flows ($Re_c = 10^3 - 10^4$), which have applications in small aerial vehicle design, are still to be explored [23]. When used for low Reynolds number ($Re_c < 5 \times 10^5$) scenarios, the performance of several airfoils having a strong performance at high Reynolds number ($Re_c = 10^6 - 10^7$) may degrade [24] [25]. The root of this deterioration may be the laminar separation bubble arising on the top of the airfoil and the corresponding bursting of the bubble [26] [27] [28] [29]. A significant number of experiments have been carried out in the past few years on the laminar separation bubble, which has sparked further work into the comprehension of the mechanics underlying low Reynolds number flows [30] [31] [32] [33] [34] [35] [36] [37]. Such studies, however, rely predominantly on single-element airfoils or simplistic flat-plate modeling.

High lift multi-element systems perform better at large Reynolds numbers ($Re_c = 10^6 - 10^7$) for which they were built, resulting in higher maximum lift co-efficient and delayed stall angle [7]. These high-lift systems are sadly major causes of air-frame noise. To improve aerodynamic efficiency [38] and minimize aeroacoustic noise [39], there has been a thorough study of the high-Reynolds-number flow associated with multi-element airfoils [40] [41] [42] [43] [44] [45] [46]. The low-Reynolds-number flow about this type of complex high-lift system is nevertheless still an unresolved problem. Improvement in this topic will further expand the understanding of low-Reynolds-number flow physics and potentially support the development of future high-performance low-Reynolds-number aircraft. The complicated design of a multi-element system leads to intrinsic interactions between one element's wake and the downstream element's boundary layer [47]. The present research will concentrate on these low Reynolds number interactions and add to the fundamental topic of intricate interactions among flow structures [47].

The slat wake for high Reynolds number differs significantly from the wake at low Reynolds number. Vortex shedding is observed from the slat cove shear layer in the case of high Reynolds number flows [42] [43]. The findings of particle image velocimetry experiments of Jenkins et al. [43] at $Re_c = 3.64 \times 10^6$ revealed that these shed vortices might travel through the gap area and interfere with the slat trailing edge wake. Furthermore, in the wake of the slat trailing edge, the blunt slat trailing edge and the detachment of the boundary layer from the upper surface of the slat may lead to shed spanwise vortices. Thus, these spanwise vortices describe the intricate slat wake.

Deck's [42] three-dimensional zonal-detached eddy simulation at $Re_c = 2.09 \times 10^6$ showed that the shed spanwise vortices resulting from the slat cove shear layer go through a three-dimensional transition and are then distorted by the rapid gap flow to develop vortices in the streamwise direction. The intricate slat wake originates from the interaction between these streamwise vortices and the spanwise vortices generated from the blunt slat trailing edge. No roll-up emerges from the slat cove shear layer for the low-Reynolds-number flow however [48] [2]. The volatility of the laminar slat wake in the critical Reynolds number range is one of the reasons behind spanwise vortex shedding as the experiments of Makiya et al. [48] showed for $8.4 \times 10^4 < Re_c < 2.1 \times 10^5$ and $2.1 \times 10^5 < Re_c < 5.9 \times 10^5$.

Spanwise vortices are thus predominant in the wake of the slat. Makiya et al. [48] did use a NACA23012 airfoil to create their multi-element airfoil however, so it is possible that some inconsistencies arise from the 30P30N case.

Coherent vortices dominate in the slat wakes at high and low Reynolds numbers, as per the literature review mentioned above. Such coherent vortices could interfere with the main element's boundary layer. These interactions have been thoroughly studied with simplistic geometries as well as with complex geometries. Earlier hot-wire observations and computational simulations identified the dominance of spanwise secondary vortices [49] and streamwise streaky structures [50] within the boundary layer for the simplistic setup of the circular cylinder upstream of the flat plate (discussed in Chapter 4). "One-to-one" correspondence was noted between secondary spanwise vortices and wake vortices [51]. Three-dimensional instability in secondary spanwise vortices may lead to hairpin vortices and consequently to the turbulent boundary layer.

Although advances have been made with simpler arrangements, the interactions between the wake vortices and the boundary layers still require additional study, particularly in complex engineering-related systems (e.g. multi-element airfoils).

Wang et al. [2] found that the slat wake of a 30P30N wing at $Re_c = 8.32 \times 10^3$ and $\alpha = 2° - 12°$ is governed by counter-rotating streamwise vortices also known as Görtler vortices (discussed in Chapter 4). No spanwise vortices were noticed in the gap region or near the blunt trailing edge of the slat. The authors proposed that the virtual curved wall generated by the upper boundary of the separation bubble in the slat cove plays an important role in the development of Görtler vortices. These Görtler vortices in the slat wake might interfere with the separated shear layer over the main element and lead to streaky structures inside it.

Recently, Wang et al. [3] performed additional experiments to further explore complex low Reynolds number flows over the 30P30N high-lift wing and proposed a critical interval based on Re_c between 1.27×10^4 and 1.38×10^4. Below the critical values, Görtler vortices are present in the slat wake. These Görtler vortices stay above the separated shear layer from the main element and generate streaky structures inside the separated shear layer similar to the case of $Re_c = 8.32 \times 10^3$. The instability in the separated shear layer leads to shed vortices on top of the main element. When Re_c is above the critical values, shed spanwise vortices from the slat cove dominate in the gap region and are later deformed by the impact on the slat trailing edge and the accelerated flow in the gap region. These deformed vortices later evolve into hairpin-like vortices. Wang et al. [3] also observed double secondary spanwise vortices above the main element initiated by the spanwise vortices coming from the slat cove region.

The current work is focused on numerical investigation of this complex flow physics for $Re_c = 8.32 \times 10^3, 1.27 \times 10^4$, and 1.83×10^4 using an open-source solver Nek5000 [1] based on high-order spectral element methods. The goal is to simulate the exact same flows as the experiments [2] [3] allowing validation and then to explore more of the flow physics.

Chapter 3

Methodology

This section covers the entire methodology starting from the preparation of the computational domain to the post-processing of results. The open-source code Nek5000 [1] was used to compute the flow field around a 30P30N three-element high lift wing. Nek5000 solves the incompressible Navier-Stokes equations using the high-order Spectral Element Method (SEM). Section 3.1 discusses the solver used, including spatial and temporal discretization details. Section 3.2 details the geometry and computational domain under consideration for both 2D and 3D computations. In total, six cases were simulated, as described in Section 3.3. The complex process of structured grid generation for the 30P30N three-element high lift device is discussed in Section 3.4 along with tools used. Section 3.5 describes necessary flow parameters for the computations. The initial and boundary conditions used for both 2D and 3D simulations are described in Section 3.6 along with mathematical details of the sponge forcing function applied at the outflow boundary. Finally, Section 3.7 presents information on intensive parallel computations and visualizations.

3.1 Solver

This section provides information about the numerical scheme and solver used including the discretization methodology.

3.1.1 Spectral Element Method and Nek5000

There are different methodologies to solve the Navier-Stokes equations such as the Finite Element Method (FEM), the Finite Difference Method (FDM), the Finite Volume Method (FVM), and higher order Spectral Methods (SM). The method to be chosen depends on the type of the problem under consideration and the level of accuracy desired.

The Spectral Element Method (SEM) [52] is a higher-order version of FEM. This method integrates the high-order accuracy of SM and the geometric flexibility of FEM. The method divides the domain under consideration into disjoint macro-elements, inside which the solution is approximated with high-order polynomials.

The SEM exhibits exponential convergence, and low dissipation and dispersion errors due to high-order implementation. This is important for investigating sensitive instabilities that lead to transition from laminar to turbulent flow and transition from 2D to 3D: the growth of these instabilities from minute disturbances cannot be smeared out by low-order schemes. High-order methods are thus crucial but come with the expense of intensive computational requirements.

In this work Nek5000 [1] software, developed by Argonne National Laboratory, was used as a solver: it is an open-source implementation of the high-order SEM [52]. Nek5000 solves the following incompressible Navier-Stokes equations,

$$\nabla.\mathbf{u} = 0, \tag{3.1}$$

$$\rho\left(\frac{\partial \mathbf{u}}{\partial t} + \mathbf{u}.\nabla\mathbf{u}\right) = -\nabla p + \mu\left(\nabla^2\mathbf{u}\right) + \rho\mathbf{f}, \tag{3.2}$$

where vectors are represented by bold letters. $\mathbf{u}$ is the velocity vector, p is pressure, ρ is density, μ is viscosity, and $\mathbf{f}$ is a external force or forcing function. Equation 3.2 can be written in the non-dimensional form as

$$\frac{\partial \mathbf{u}^*}{\partial t^*} + \mathbf{u}^*.\nabla \mathbf{u}^* = -\nabla p^* + \frac{1}{Re}\left(\nabla^2 \mathbf{u}^*\right) + \mathbf{f}^*, \tag{3.3}$$

where $u^* = \frac{\mathbf{u}}{V_\infty}$, $t^* = \frac{tV_\infty}{c}$, $f^* = \frac{fc}{V_\infty^2}$, $p^* = \frac{p}{\rho V_\infty^2}$, and $Re = \frac{\rho V_\infty c}{\mu}$.

3.1.2 Spatial Discretization

The data is represented on sets of non-overlapping sub-domains, Ω^n, within the entire domain Ω_f.

$$\Omega_f = \bigcup_{n=1}^{N_f} \Omega^n; n = 1, ..., N_f. \tag{3.4}$$

Any given variable, $\phi(x)$, inside each element Ω^n is represented by

$$\phi(\mathbf{x})|_{\Omega^n} = \phi(x,y,z)|_{\Omega^n} = \sum_{k=0}^{N}\sum_{j=0}^{N}\sum_{i=0}^{N} \phi_{ijk}^n h_i(r)h_j(s)h_k(t), \tag{3.5}$$

where $h_i(r)$, $h_j(r)$, $h_k(r)$ are N^{th} order one-dimensional Lagrangian interpolant Legendre polynomials based on the Gauss Lobatto Legendre quadrature points, ϕ_{ijk}^n are the basis coefficients (the values of ϕ at the quadrature points), and $\mathbf{r} = (r,s,t)$ are the coordinates in the canonical reference element $\hat{\Omega} = [-1,1]^3$. For $\mathbf{x} = (x,y,z) \in \Omega^n$, an iso-parametric mapping is given by

$$\mathbf{x}|_{\Omega^n} = \sum_{k=0}^{N}\sum_{j=0}^{N}\sum_{i=0}^{N} \mathbf{x}_{ijk}^n h_i(r)h_j(s)h_k(t). \tag{3.6}$$

In a similar manner discretization of the pressure space can be performed, however, Lagrangian interpolants of order $N - 2$ are used. This is commonly known as $\mathbb{P}_N - \mathbb{P}_{N-2}$ staggering. For example, if the order of polynomial for velocity is $N = 10$ then for pressure it is $N = 8$. This formulation is very useful to remove the occurrence of unwanted (spurious) pressure modes that can arise in the Navier-Stokes equations and the requirement of pressure boundary conditions [53].

The governing equations are written in the weighted residual form, and the test functions are used to minimize the residuals across the entire domain. The spectral element method ensures continuity at elemental boundaries and weak continuity in the first derivatives at those boundaries. As the polynomial order increases and the resolution becomes better, the smoothness of the solutions at element boundaries increases. When the resolution is insufficient, noise (in the form of oscillations) appears at the element boundaries. As the Reynolds number increases to large values $(O(10^5 - 10^6))$, this becomes a significant problem. The elemental integrals of the weak formulation of the Navier-Stokes equations are evaluated by Gauss-Lobatto quadrature, yielding matrix equations of the form

$$A^k \phi^k = B^k f^k, \tag{3.7}$$

where A^k is an elemental stiffness matrix, B^k is an elemental mass matrix, ϕ^k is any given variable on the element Ω^k, and f^k is the right-hand side of the velocity or pressure Poisson equations. The elemental residuals are added at element boundaries in a direct stiffness summation procedure of assembling elemental matrix blocks. For more details on this entire procedure see Deville et al. [54].

3.1.3 Temporal Discretization

Nek5000 implements a semi-implicit time-stepping method. The viscous terms are treated implicitly using a k-step backward differentiation formula (BDFk) whereas the non-linear terms are treated with an explicit extrapolation scheme of order k (EXTk). For the cases studied in this work, a third-order accurate BDF3/EXT3 time-stepping scheme was used.

A restriction exists on the step size Δt because of the explicit formulation of the convective terms to maintain global stability during time-marching. This stability is governed by the Courant Friedrichs Lewy (CFL) condition number, which can be written as

$$\Delta t < C \times min_{\Omega_f}\left(\frac{\Delta x}{|u|}, \frac{\Delta y}{|v|}, \frac{\Delta z}{|w|}\right), \tag{3.8}$$

where C represents CFL number, min_{Ω_f} indicates minimum over the entire flow field, and $\Delta x, \Delta y, \Delta z$ are the x, y, z distances between the collocation points.

3.1.4 Time Splitting Scheme and General Solution Procedure

The semi-discrete formulation described in Karniadakis et al. [55] is briefly discussed in this section. We rewrite Equation 3.3 (dropping $*$ notation) as

$$\frac{\partial \mathbf{u}}{\partial t} = -\nabla p + \mathbf{N}(\mathbf{u}) + C_1 \mathbf{L}(\mathbf{u}) + f, \tag{3.9}$$

where vectors are represented by bold letters. Here, $\mathbf{N}$ and $\mathbf{L}$ represent the nonlinear and linear operators, respectively, defined as

$$\mathbf{N}(\mathbf{u}) = -\mathbf{u}.\nabla\mathbf{u}, \tag{3.10}$$

$$\mathbf{L}(\mathbf{u}) = \nabla^2\mathbf{u}. \tag{3.11}$$

Further, Equation 3.9 can be split into three substeps as

$$\widehat{\mathbf{u}} = \sum_{j=1}^{k} a_j \mathbf{u}^{n+1-j} + \Delta t\left(\sum_{j=1}^{k} b_j \mathbf{N}(\mathbf{u}^{n+1-j}) + \mathbf{f}^{n+1}\right), \tag{3.12}$$

$$\widehat{\widehat{\mathbf{u}}} = \widehat{\mathbf{u}} - \Delta t\nabla p^{n+1}, \tag{3.13}$$

$$\mathbf{u}^{n+1} = \widehat{\widehat{\mathbf{u}}} + \Delta t C_1 \mathbf{L}(\mathbf{u}^{n+1}). \tag{3.14}$$

Here $\widehat{\mathbf{u}}$ and $\widehat{\widehat{\mathbf{u}}}$ are intermediate velocity fields; a_j and b_j refer to the k coefficients of k-step backward differentiation formula (BDFk) and the k coefficients of the explicit extrapolation scheme of order k (EXTk), respectively. Finally, an elliptic equation (Poisson equation) for the pressure can be written as

$$\nabla^2 p^{n+1} = \frac{1}{\Delta t} \nabla . \widehat{\mathbf{u}}. \tag{3.15}$$

The final velocity field $\mathbf{u}^{n+1}$ is obtained by solving the Helmholtz Equation 3.14 where velocity field $\widehat{\widehat{\mathbf{u}}}$ acts as a forcing term.

3.2 Geometry and Computational Domain

We consider the 30P30N three-element high lift airfoil shown in Figure 3.1. It is composed of three elements: slat, main, and flap. The stowed chord length (c) of the airfoil is 1m in the presented simulations. The Reynolds number is always based on stowed chord length in this work. The slat and flap chord lengths are 15% and 30% of the stowed chord length, respectively. Other geometrical parameters related to the airfoil are stated in Table 3.1.

Slat Angle, δ_s	30°
Flap Angle, δ_f	30°
Slat Gap, g_s	2.95%
Flap Gap, g_f	1.27%
Slat Overhang, o_s	-2.5%
Flap Overhang, o_f	0.25%

Table 3.1: Geometrical parameters for 30P30N three-element high lift airfoil [56]

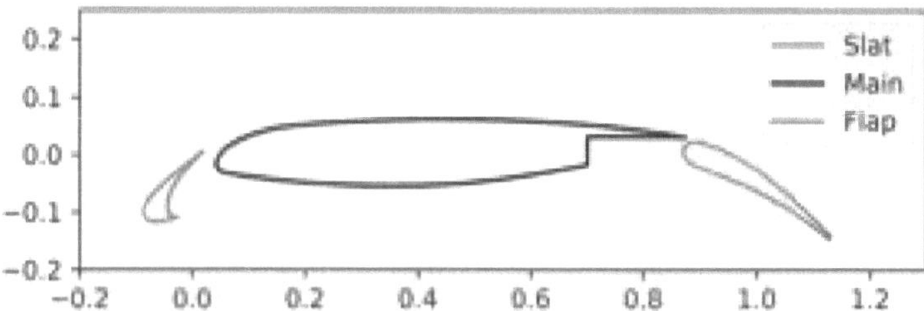

Figure 3.1: 30P30N three-element high lift airfoil

3.2.1 2D Case

The 2D computational domain is in the shape of a rectangle, with 3c and 4.75c lengths upstream and downstream of the airfoil respectively. The rectangular domain is used to match the rectangular geometry of the water tunnel used in the experiments [2] [3]. The domain is 9c long (in X, streamwise direction) and 5c (in Y) tall which is in accordance with the experimental domain mentioned in [2] [3]. Different sizes of domains, both larger and smaller, were considered, and the final domain (Figure 3.2) was deemed sufficient for the computation without compromising on the solution quality. The boundary conditions (BCs) used are indicated in Figure 3.2 with the computational domain. The boundary conditions are discussed in detail in Section 3.6.

3.2.2 3D Case

The 3D computational domain has the same shape and dimensions as the 2D computational domain with a spanwise length of 0.2c (in Z, spanwise direction). The spanwise length was chosen based on expected wavelengths of vortical structures and previous DNS calculations performed on a NACA 4412 wing by Hosseini et al. [57] using Nek5000. The periodic boundary condition is applied in the spanwise direction. Although we define the domain as a 3D computational domain,

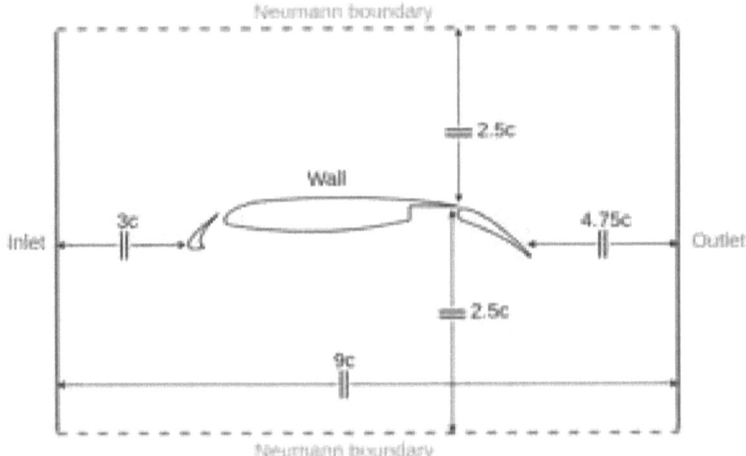

Figure 3.2: Computational domain with boundary conditions for the 30P30N three-element airfoil (c = stowed chord length)

it is typically known as 2.5D or a 3D slice because the airfoil is simply extruded in the spanwise direction (without sweep) to allow the development of 3D flow structures.

3.3 Cases Simulated

Several cases are simulated to study the effect of Reynolds number and dimensionality on the flow field. The parameters for these cases are given in Table 3.2 in terms of dimensionality, Reynolds number and angle of attack (AOA).

Case	Dimensionality	Re_c	AOA
Case 1	2D	8.32×10^3	$4°$
Case 2	2D	1.27×10^4	$4°$
Case 3	2D	1.83×10^4	$4°$
Case 4	3D	8.32×10^3	$4°$
Case 5	3D	1.27×10^4	$4°$
Case 6	3D	1.83×10^4	$4°$

Table 3.2: Cases under consideration

3.4 Grid Generation

The geometry and mesh are generated using the open-source software Gmsh [58]. As Nek5000 supports only a structured grid with quadrilateral and hexahedral elements, the domain under study consists of only quadrilateral and hexahedral elements for 2D and 3D simulations respectively. A multi-block grid strategy is applied to generate a structured grid that has 49 blocks as shown in Figure 3.3. The gmsh2nek [59] converter was used to export the mesh in Nek5000 suitable format. Also, mesh quality is examined using checkMesh, a built-in mesh quality checker in OpenFoam [60]. Different parameters are considered such as skewness, orthogonality, and aspect ratio to ensure the mesh quality of preliminary meshes generated using gmsh before converting to the Nek5000 supported format. In addition, a mesh smoother [61] specifically written for Nek5000 computation was used in run-time to further smooth out the mesh.

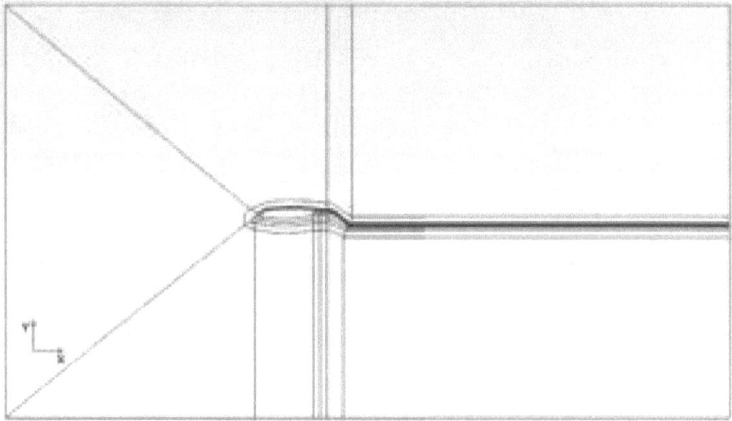

Figure 3.3: Multi-block grid strategy inside rectangular domain for the 30P30N three-element airfoil

The 2D mesh consists of a total of 13,282 elements in the XY plane. A 2D-close-up of the spectral element mesh in the XY plane is shown in Figure 3.4. Figure 3.5 illustrates the entire spectral element grid. The polynomial degree is set to $N = 7$ for velocity in accordance with the grid independence study discussed in Section 4.1.1 which results in a total of 650,818 spectral collocation points. The density of elements is high near the surface of the airfoil to resolve the boundary layer properly, and decreases progressively towards the farfield. Increasing the density of elements below the main airfoil is difficult with the present multi-block approach. Nevertheless, solutions with higher grid density below the main airfoil were computed, yet showed no significant changes. However, more elements are placed in the slat cove region and on top of the main element as these areas are of primary interest.

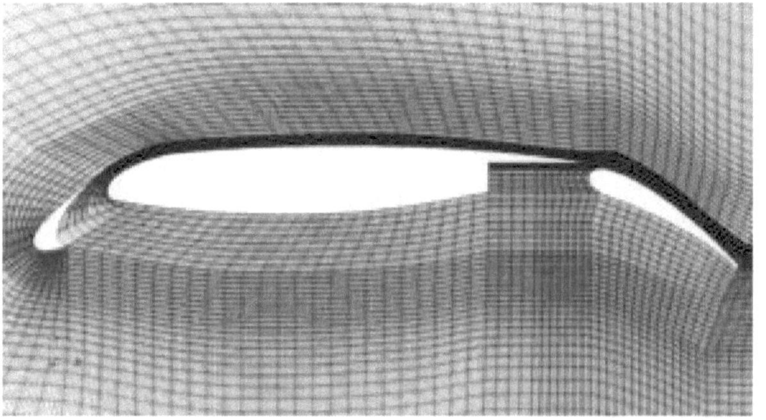

Figure 3.4: 2D-close-up of the spectral element grid for the 30P30N three-element airfoil

The 3D mesh has exactly the same topology and the same number of elements in the XY plane as the 2D mesh. The 3D mesh is generated by simply extruding the 2D mesh in the spanwise direction. In addition, the spanwise direction is resolved using ten equally-sized elements with an order of polynomial $N = 7$ for velocity, resulting in a total of 132,820 hexahedral elements and 46 million spectral collocation points for the 3D mesh of 20% span. Close-ups of the 2D spectral element grid for the three separate elements of the airfoil are shown in Figure 3.6.

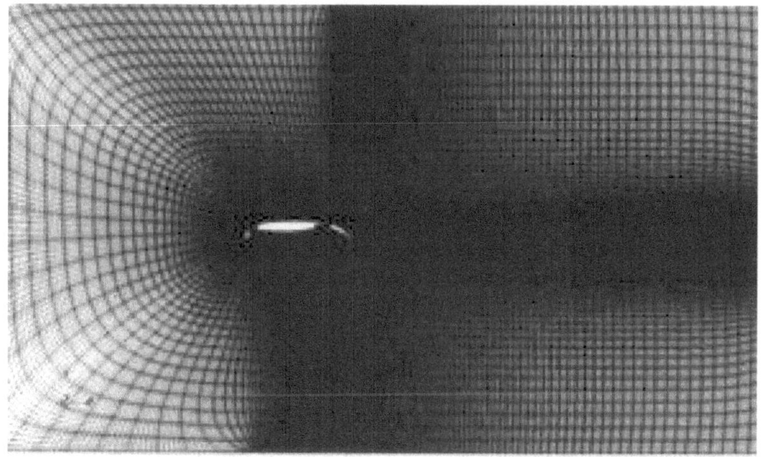

Figure 3.5: 2D spectral element grid for the 30P30N three-element high lift airfoil (13,282 quadrilateral elements and 650 thousand spectral collocation points)

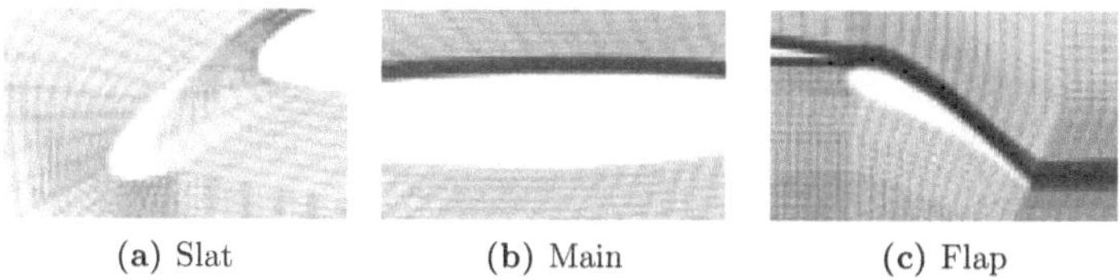

(a) Slat   (b) Main   (c) Flap

Figure 3.6: 2D close-ups of the 2D spectral element grid around the slat (a), main element (b), and flap (c) for the 30P30N three-element airfoil

3.5 Flow Parameters

As the simulations carried out are based on the incompressible Navier-Stokes equations (Section 3.1), the only similarity parameter to be considered is the stowed-chord Reynolds Number (Re_c), which is given by,

$$Re_c = \frac{\rho V_\infty c}{\mu},$$ (3.16)

where c is unit stowed-chord length, ρ is unit density, and μ represents dynamic viscosity. Free stream velocity, V_∞, is set to one. The value of dynamic viscosity is altered to study the flow at different Reynolds numbers. Therefore, the dynamic viscosity is denoted as a function of Reynolds number,

$$\mu = \frac{1}{Re_c}.$$

(3.17)

3.6 Initial and Boundary Conditions

All the 2D and 3D computational domains are uniformly initialized with free-stream velocity (unit value throughout the domain).

Figure 3.2 describes the boundary conditions used for presented computations. For the 2D computations, a constant velocity boundary condition (v) is used at the inlet with a unit value. A wall boundary condition (W) (zero-slip) is applied on the surface of the airfoil. The outflow or open boundary condition (O) is applied on the outlet. This outflow boundary condition stems from the incompressible Navier-Stokes variational formulation as a natural boundary condition. In the present case, the no-stress formulation of the outflow or open boundary condition is given by,

$$[-p\mathbf{I} + \mu(\nabla \mathbf{u})].\mathbf{n} = 0$$

(3.18)

where $\mathbf{n}$ is the unit vector, $\mathbf{I}$ is the identity tensor, and μ is dynamic viscosity. This outflow boundary is dampened by implementing a sponge forcing function as described in Section 3.6.1. The Neumann boundary condition is imposed on the top and bottom of the computation domain for all velocity components given by,

$$\frac{\partial \mathbf{u}}{\partial y} = 0.$$

(3.19)

For the 3D computations, all boundary conditions are the same as for the 2D computation except an additional periodic boundary condition (P) is applied in the spanwise direction.

3.6.1 Sponge Forcing

A sponge forcing function is applied at the outflow boundary condition due to the relatively small distance between the trailing edge of the flap and the outflow boundary, which may lead to back-flow at the outflow boundary. This forcing function was originally developed by Schrader et al. [62] [63]. This function restrains the flow to a referred velocity. In the presented computations, this sponge forcing function was applied over a region of 10% of stowed chord length before the end of the domain (outflow) in the streamwise direction. This is achieved by inserting the sponge forcing term on the right-hand side of Equation 3.2 given by

$$\mathbf{F}(x,t) = A_{max}\lambda_f(x)[\mathbf{U}_f(x) - \mathbf{U}(x,t)], \tag{3.20}$$

where $\mathbf{U}_f$ indicates the forcing velocity (unit value in presented simulations) and $\mathbf{U}$ indicates the instantaneous velocity. The λ_f is a step function given by

$$\lambda_f(x) = \left[S\left(\frac{x - h_{start}}{h_{rise}}\right) - S\left(\frac{x - h_{end}}{h_{fall}} + 1\right) \right], \tag{3.21}$$

and S is given by

$$S(\xi) = \begin{cases} 0 & \xi \leq 0 \\ 1/(1 + e^{(1/(\xi-1)+1/\xi)}) & 0 < \xi < 1 \\ 1 & \xi \geq 1. \end{cases} \tag{3.22}$$

The A_{max} was chosen to be 1. At this value, no pressure waves were noted to be coming back from the outflow boundary up to $Re_c = 1.83 \times 10^4$. The h_{start} and h_{end} indicate the starting and ending locations of the sponge forcing zone. The selected values are

$$h_{rise} = 0.6(h_{end} - h_{start}) \qquad (3.23)$$

$$h_{fall} = 0.3(h_{end} - h_{start}). \qquad (3.24)$$

3.7 Parallel Computation and Visualization

The computation for all the cases was carried out on Compute Canada clusters on 16 nodes with 32 cores each, resulting in a total of 512 cores. This is made possible by MPI implementation in Nek5000 [1]. Nek5000 is highly scalable and has been successfully tested on petascale systems by Offermans et al. [64] with excellent performance on up to one million cores. Table 3.3 indicates the time taken per flow-through (the computational domain) for each case. In total, four terabytes of data were generated during computations.

Case	Dimensionality	Re_c	AOA	Wallclock time per flow-through (hr)
Case 1	2D	8.32×10^3	$4°$	1.5
Case 2	2D	1.27×10^4	$4°$	1.8
Case 3	2D	1.83×10^4	$4°$	2.0
Case 4	3D	8.32×10^3	$4°$	67.8
Case 5	3D	1.27×10^4	$4°$	71.3
Case 6	3D	1.83×10^4	$4°$	77.2

Table 3.3: Computation time

Paraview [65] [66], an open-source visualization software, was used to visualize results and render images to understand the flow physics. The massive post-processing was made possible by using Paraview in parallel on 128 cores.

Chapter 4

Results and Analysis

In this section, results achieved from spectral element simulations of flow past a 30P30N airfoil (2D) and wing (3D) at the three Reynolds numbers ($Re_c = 8.32 \times 10^3$, 1.27×10^4, and 1.83×10^4) studied experimentally by Wang et al. [2] [3] are presented. The drag and lift measurements along with a grid independence study are presented in Section 4.1 for 2D computations at the three Reynolds numbers mentioned above. Section 4.2 documents the pressure coefficient plots achieved from 2D simulations at the three Reynolds numbers mentioned above. The contours of vorticity for both 2D and 3D simulations at the three Reynolds numbers mentioned above are presented in Section 4.3 including comparison with experimental data. Section 4.4 details the three-dimensional vortical structures, similar to mode A ($Re_c = 8.32 \times 10^3$) and mode B ($Re_c = 1.27 \times 10^4$ and 1.83×10^4) instabilities, observed in the present simulations along with their implications for the flow. The Görtler vortices observed in the case of $Re_c = 8.32 \times 10^3$ and 1.27×10^4 are described in Section 4.5. Finally, Section 4.6 discusses the hairpin vortices observed in the case of $Re_c = 1.83 \times 10^4$.

4.1 Drag and Lift Measurements

4.1.1 Grid Independence

A grid independence study is a very crucial component of any numerical simulation. In simple terms, it means no substantial change is observed in the flow behaviour as we increase the grid resolution. In Nek5000 [1], the grid is composed of a large number of quadrilateral (in 2D) and hexahedral (in 3D) rectilinear or curved elements inside which the solution is modeled as a tensor product of Nth order Legendre polynomials (N in each direction). Once the elemental mesh is found to be adequate to capture the flow, increasing polynomial order N serves as a way to further increase resolution. $N + 1$ is the number of collocation points in each direction in each element. Tables 4.1 and 4.2 show the variation in the time-averaged drag values and time-averaged lift values, respectively, for the case of $Re_c = 1.83 \times 10^4$ at $AOA = 4°$ as we move towards high resolution ($N + 1 = 12$) from low resolution ($N + 1 = 8$). No substantial change is observed in the value of time-averaged drag coefficient (% change $< 0.22\%$, Table 4.1, $\overline{C_D}$,) and time-averaged lift coefficient (% change $< 1.33\%$, Table 4.2, $\overline{C_L}$,) over a calculation of over three flow-throughs (also referred to as flow passes) of the domain. This confirms the grid independence of the simulation and implies that the resolution of $N + 1 = 8$ is adequate to capture the flow characteristics. Also, 3.3 flow passes is a low number for unsteady lift and drag coefficients to converge as can be seen by the instantaneous time traces (Figures 4.2, 4.3, 4.4, 4.6, 4.7, 4.8) but in order to save computation time this grid convergence study was performed and deemed adequate at 1% change. Similar grid convergence studies were performed for lower Reynolds numbers (not shown here for conciseness).

N+1	Order of polynomial	$\overline{C_D}$	% Change	Flow passes
8	7	0.84334	-	3.3
10	9	0.84420	0.102	3.3
12	11	0.84516	0.216	3.3

Table 4.1: Grid convergence indicated by time-averaged values of drag coefficients for different orders, N, of polynomials for $Re_c = 1.83 \times 10^4$ (2D case)

N+1	Order of polynomial	$\overline{C_L}$	% Change	Flow passes
8	7	1.39948	-	3.3
10	9	1.38891	0.755	3.3
12	11	1.38090	1.328	3.3

Table 4.2: Grid convergence indicated by time-averaged values of lift coefficients for different orders, N, of polynomials for $Re_c = 1.83 \times 10^4$ (2D case)

4.1.2 Drag Coefficient

Figure 4.1 shows the change in the value of time-averaged drag coefficient with Reynolds number. The time-averaged drag coefficient decreases with increase in Reynolds number in this particular range, unlike the experiments performed at high Reynolds numbers ($Re_c > 1 \times 10^6$) [9] [10] [14] [67]. One of the reasons can be the laminar separation bubble observed on top of the main element in the case of $Re_c = 8.32 \times 10^3$ and $Re_c = 1.27 \times 10^4$, that is bigger in case of the former (Figures 4.15, 4.16, 4.20, and 4.21). This is not the case for $Re_c = 1.83 \times 10^4$ as no laminar separation bubble is observed on top of the main element in this case (Figures 4.25 and 4.26).

Figures 4.2, 4.3 and 4.4 indicate the time series of drag coefficient along with the time-averaged value for $Re_c = 8.32 \times 10^3$, $Re_c = 1.27 \times 10^4$, and $Re_c = 1.83 \times 10^4$ respectively. Averaging was performed over a period of 8.3 flow passes.

4.1.3 Lift Coefficient

Figure 4.5 shows the change in the value of time-averaged lift coefficient with Reynolds number. The time-averaged lift coefficient increases with increase in Reynolds number which matches the different experiments performed in the past

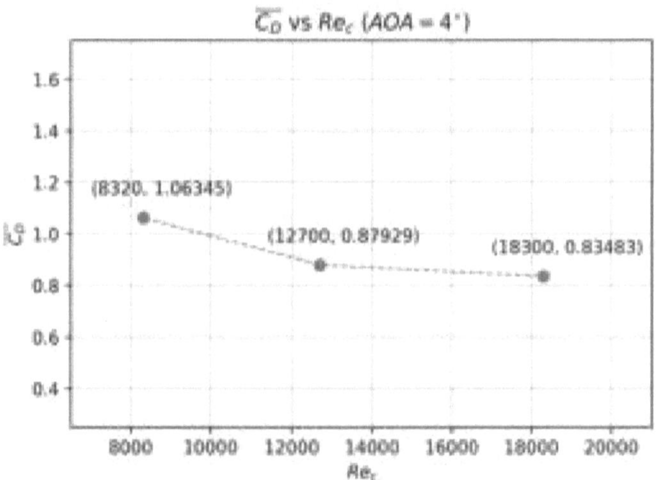

Figure 4.1: Variation of time-averaged drag coefficient with Reynolds number for flow around the 30P30N airfoil at AOA = 4° (2D case)

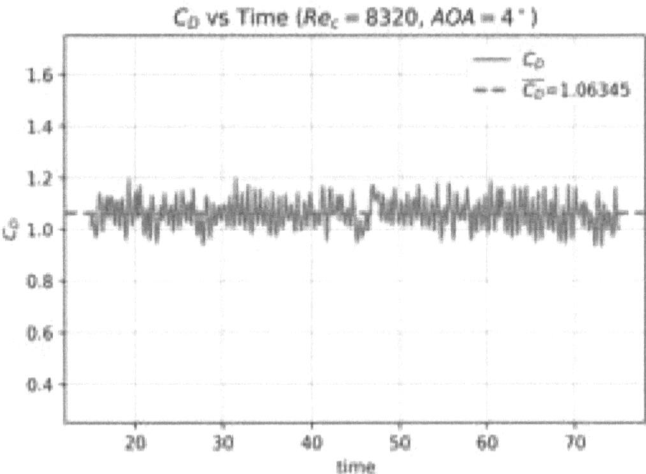

Figure 4.2: Drag coefficient variation with time for flow around the 30P30N airfoil for $Re_c = 8.32 \times 10^3$ and AOA = 4° (2D case)

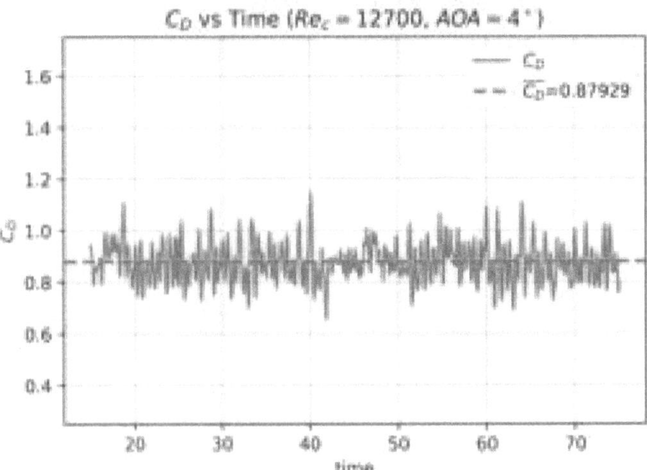

Figure 4.3: Drag coefficient variation with time for flow around the 30P30N airfoil for $Re_c = 1.27 \times 10^4$ and AOA = 4° (2D case)

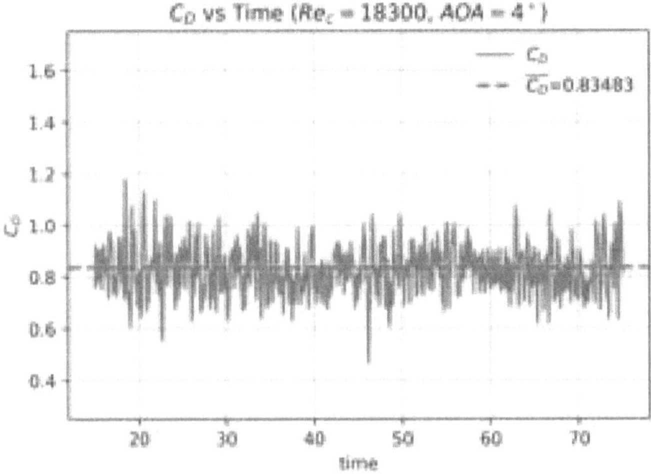

Figure 4.4: Drag coefficient variation with time for flow around the 30P30N airfoil for $Re_c = 1.83 \times 10^4$ and AOA = 4° (2D case)

[9] [10] [14] [67]. Also, this can be confirmed by measuring areas under the pressure coefficient curves (Figures 4.9, 4.10, and 4.11).

Figures 4.6, 4.7 and 4.8 indicate the time series of lift coefficient along with the time-averaged value for $Re_c = 8.32 \times 10^3$, $Re_c = 1.27 \times 10^4$, and $Re_c = 1.83 \times 10^4$ respectively. Averaging was performed over a period of 8.3 flow passes.

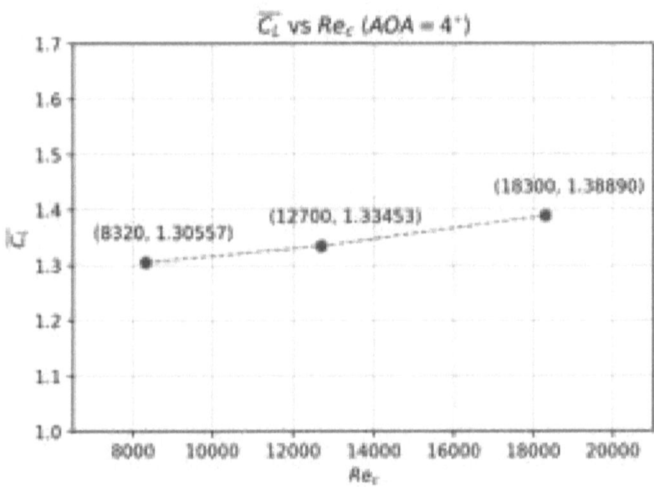

Figure 4.5: Variation of time-averaged lift coefficient with Reynolds number for flow around the 30P30N airfoil at AOA = 4° (2D case)

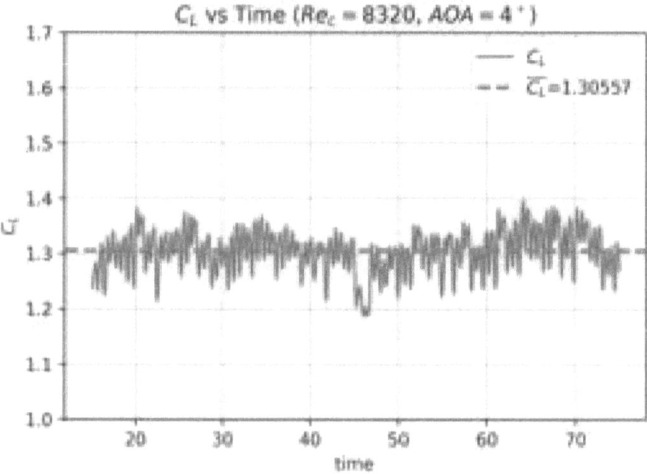

Figure 4.6: Lift coefficient variation with time for flow around the 30P30N airfoil for $Re_c = 8.32 \times 10^3$ and AOA = 4° (2D case)

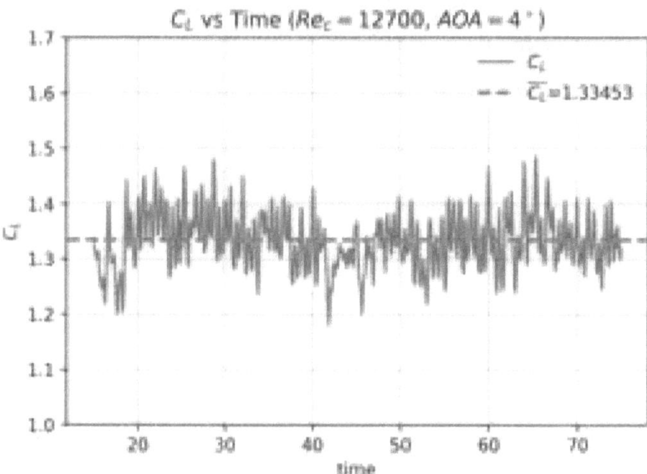

Figure 4.7: Lift coefficient variation with time for flow around the 30P30N airfoil for $Re_c = 1.27 \times 10^4$ and AOA = 4° (2D case)

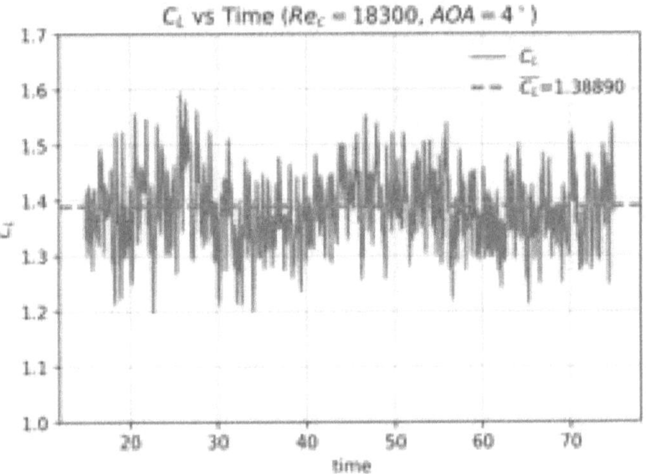

Figure 4.8: Lift coefficient variation with time for flow around the 30P30N airfoil for $Re_c = 1.83 \times 10^4$ and AOA $= 4°$ (2D case)

4.2   Pressure Coefficients

Figures 4.9, 4.10, and 4.11 represent the variation of pressure coefficient along the airfoil surface for the three Reynolds numbers considered (2D cases). The overall trend of the pressure coefficient curves described in this section for low-Reynolds numbers matches well with that of high-Reynolds numbers [14] [67]. However, the variation of pressure coefficient on the surface of the slat for the low-Reynolds number is higher than for high-Reynolds numbers at the same AOA [14] [67]. The pressure on the suction side of the main element increases noticeably with an increase in the Reynolds number. This results in an overall increase of area under the pressure coefficient curve as we move towards a higher Reynolds number, which leads to high lift values at higher Reynolds numbers. Similarly, for the slat and flap the suction peak rises with Reynolds number while the overall shape of the curves remains the same.

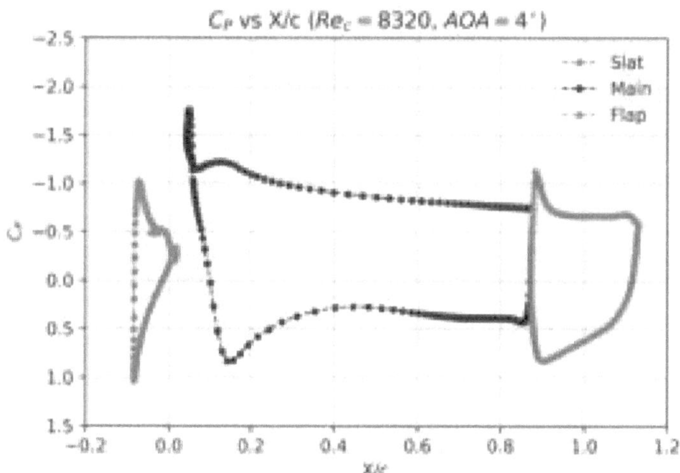

Figure 4.9: Pressure coefficient variation along the airfoil surface for flow around the 30P30N airfoil for $Re_c = 8.32 \times 10^3$ and AOA = 4° (2D case)

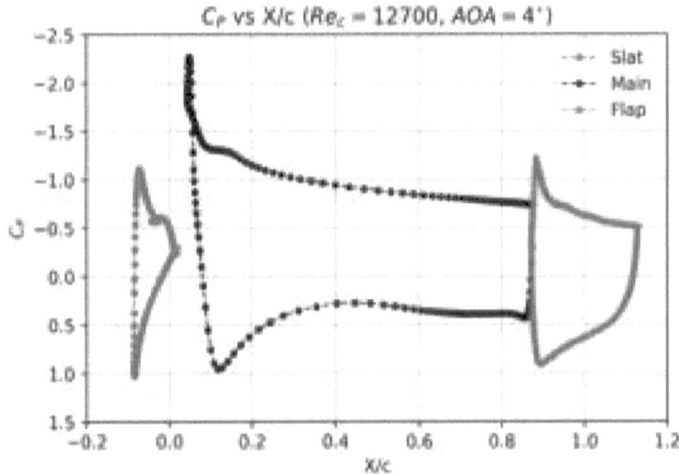

Figure 4.10: Pressure coefficient variation along the airfoil surface for flow around the 30P30N airfoil for $Re_c = 1.27 \times 10^4$ and AOA = 4° (2D case)

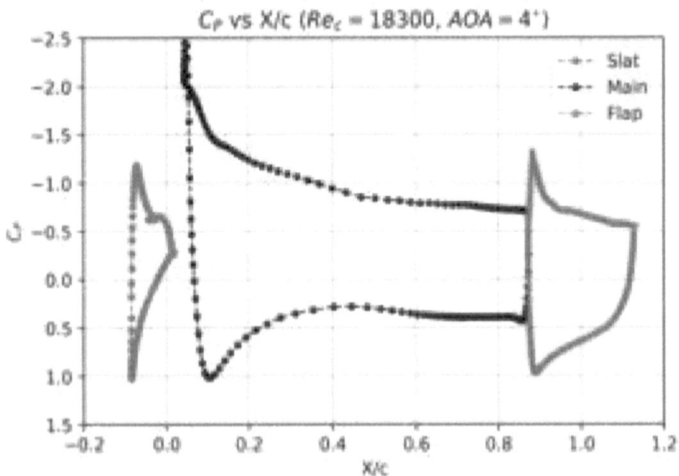

Figure 4.11: Pressure coefficient variation along the airfoil surface for flow around the 30P30N airfoil for $\hat{Re}_c = 1.83 \times 10^4$ and AOA = 4° (2D case)

4.3 Vorticity Contours

This section describes Z-vorticity contours obtained for both 2D and 3D simulations for $Re_c = 8.32 \times 10^3$, 1.27×10^4, and 1.83×10^4. The comparison with available experimental data is also presented. Additional plots of velocity and pressure contours are available in Appendix A and B, respectively.

4.3.1 $Re_c = 8.32 \times 10^3$

Figures 4.12, 4.13, and 4.14 represent the instantaneous vorticity contours for 2D (Figure 4.12) and 3D (Figures 4.13, 4.14) simulations for $Re_c = 8.32 \times 10^3$. Figures 4.15 and 4.16 represent time-averaged vorticity contours for 2D and 3D, respectively.

In both 2D and 3D cases, a shear layer labeled "Shear layer 1 (slat)" (Figures 4.12, 4.13, 4.14, 4.15, 4.16) forms on the bottom surface of the slat and persists in the gap area. No roll-up of "Shear layer 1 (slat)" is observed in this case. A shear layer labeled "Shear layer 2 (slat)" forms on the top surface of the slat and might interact with the "Shear layer (main element)", a third shear layer associated with the main element, on top of the main element. This "Shear layer (main element)" forms due to flow separation at the leading edge of the main element. A separation bubble labeled "Separation bubble (slat)" forms in the slat cove region similarly to what one would expect in flow in a cavity [68] [69] [70] or backward-facing step [71] [72] case (Figures 4.12, 4.13, 4.15).

At a later time, in 3D, the roll-up and subsequent shed spanwise vortices are observed coming from the "Shear layer (main element)" on top of the main element (Figure 4.14). This is due to a Kelvin–Helmholtz instability of the "Shear layer (main element)", coming off the main element. These shed spanwise vortices are not observed within the gap area, suggesting the laminar reattachment of "Shear layer 1 (slat)" and formation of a laminar "Separation bubble (slat)" in the slat cove [72] (Figures 4.12, 4.13, 4.15). However, a distortion is observed in the

"Separation bubble (slat)" at a later time (Figure 4.14). These results match well with experimental data [2].

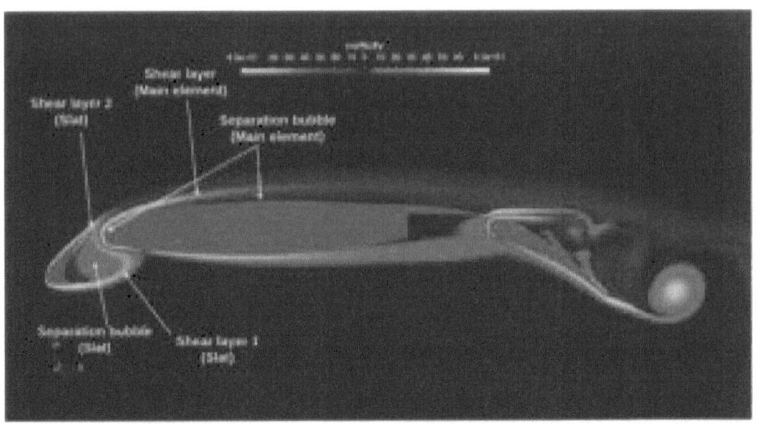

Figure 4.12: Instantaneous Z-vorticity contours for flow around the 30P30N airfoil for $Re_c = 8.32 \times 10^3$ and AOA = 4° at $tU_\infty/c = 69.80$ (2D case)

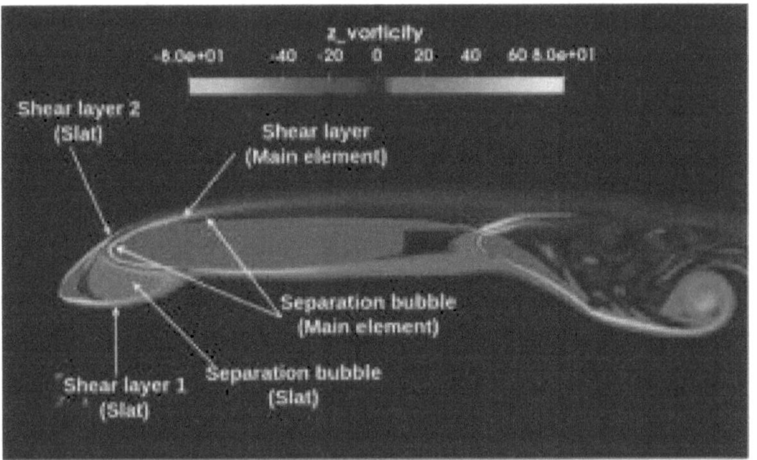

Figure 4.13: Instantaneous Z-vorticity contours for flow around the 30P30N wing for $Re_c = 8.32 \times 10^3$ and AOA = 4° at midspan and $tU_\infty/c = 13.00$ (3D case)

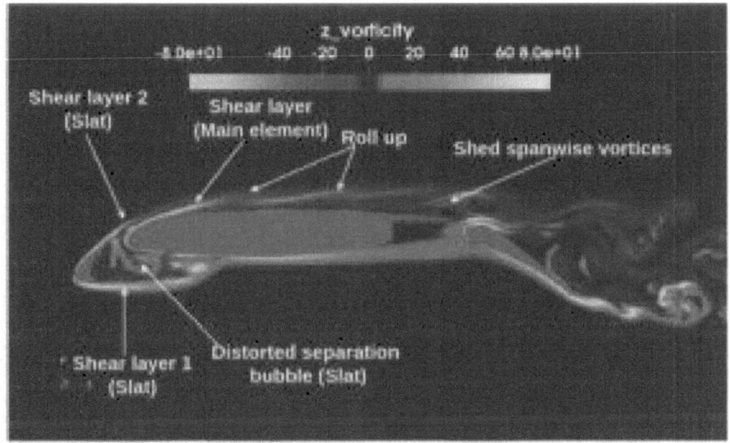

Figure 4.14: Instantaneous Z-vorticity contours for flow around the 30P30N wing for $Re_c = 8.32 \times 10^3$ and AOA = 4° at midspan and $tU_\infty/c = 18.04$ (3D case)

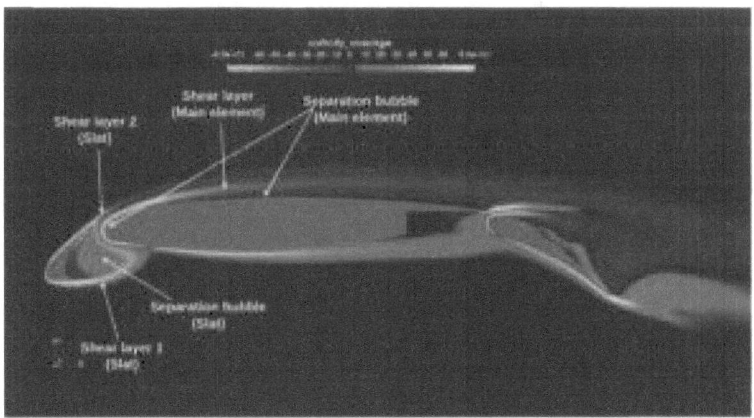

Figure 4.15: Time averaged Z-vorticity contours for flow around the 30P30N airfoil for $Re_c = 8.32 \times 10^3$ and AOA = 4° (2D case)

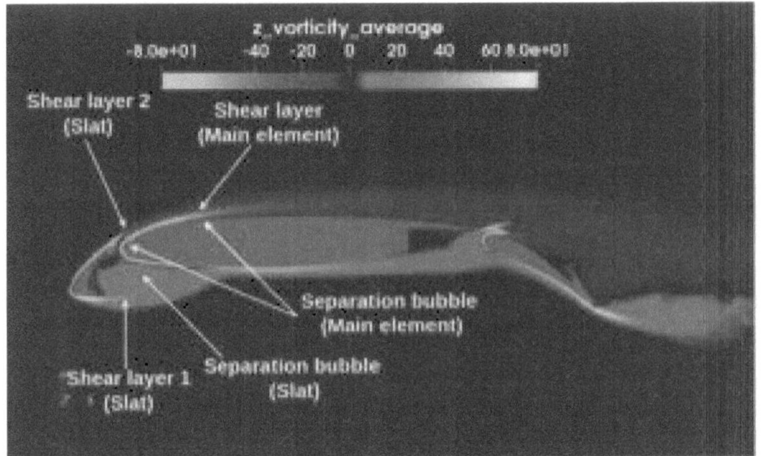

Figure 4.16: Time averaged Z-vorticity contours for flow around the 30P30N wing for $Re_c = 8.32 \times 10^3$ and AOA = $4°$ at midspan (3D case)

4.3.2 $Re_c = 1.27 \times 10^4$

Figures 4.17, 4.18, and 4.19 represent the instantaneous vorticity contours for 2D (Figure 4.17) and 3D (Figures 4.18, 4.19) simulations for $Re_c = 1.27 \times 10^4$. Figures 4.20 and 4.21 represent time-averaged vorticity contours for 2D and 3D, respectively.

In both 2D and 3D cases, a shear layer labeled "Shear layer (slat)" (Figures 4.17, 4.18, 4.19, 4.20, 4.21) forms on the leading edge of the slat and persists in the gap area. As for $Re_c = 8.32 \times 10^3$, no roll-up is observed in the slat cove from the "Shear layer (slat)". A second shear layer associated with the main element labeled "Shear layer (main element)" forms on top of the main element due to flow separation at the leading edge of the main element. This leads to formation of a separation bubble labeled "Separation bubble (main element)" on top of the main element (Figures 4.17, 4.18, 4.20).

At a later time, in 3D, the roll-up and following spanwise shed vortices are noticed downstream on the main element due to a Kelvin–Helmholtz instability of the "Shear layer (main element)", coming off the main element (Figures 4.18, 4.19). Also, the "Shear layer (slat)" coming from the leading edge of the slat reattaches to the trailing edge of the slat in a laminar state generating a laminar separation bubble in the slat cove region. No spanwise vortex shedding is observed in the slat cove (Figures 4.17, 4.18, 4.20). A distortion is observed in the "Separation bubble (slat)" at later time (Figure 4.19) that is similar to the case of $Re_c = 8.32 \times 10^3$ (Figure 4.14). These results are in harmony with the experimental results described for $Re_c = 1.27 \times 10^4$ [3].

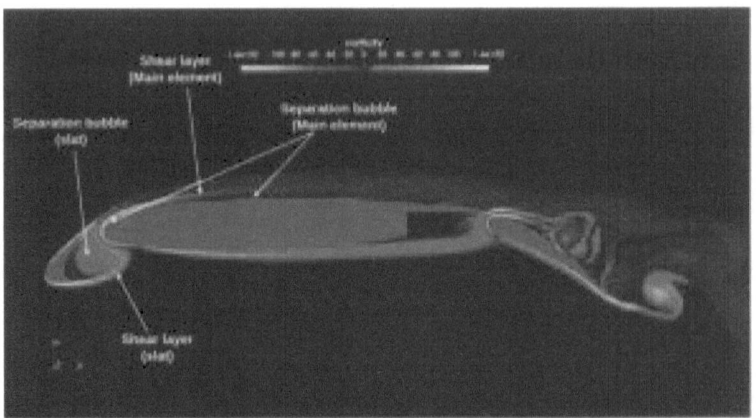

Figure 4.17: Instantaneous Z-vorticity contours for flow around the 30P30N airfoil for $Re_c = 1.27 \times 10^4$ and AOA = 4° at $tU_\infty/c = 74.90$ (2D case)

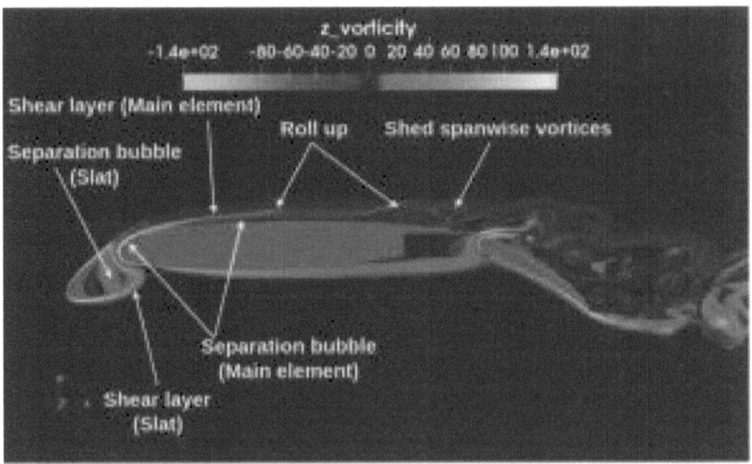

Figure 4.18: Instantaneous Z-vorticity contours for flow around the 30P30N wing for $Re_c = 1.27 \times 10^4$ and AOA = 4° at midspan and $tU_\infty/c = 9.00$ (3D case)

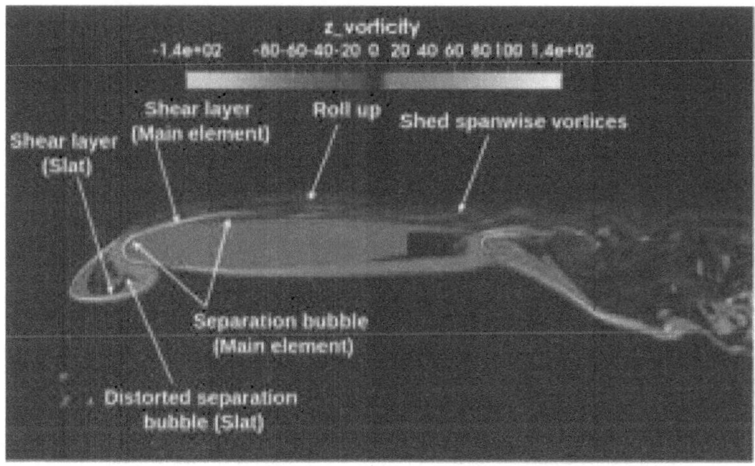

Figure 4.19: Instantaneous Z-vorticity contours for flow around the 30P30N wing for $Re_c = 1.27 \times 10^4$ and AOA = 4° at midspan and $tU_\infty/c = 11.27$ (3D case)

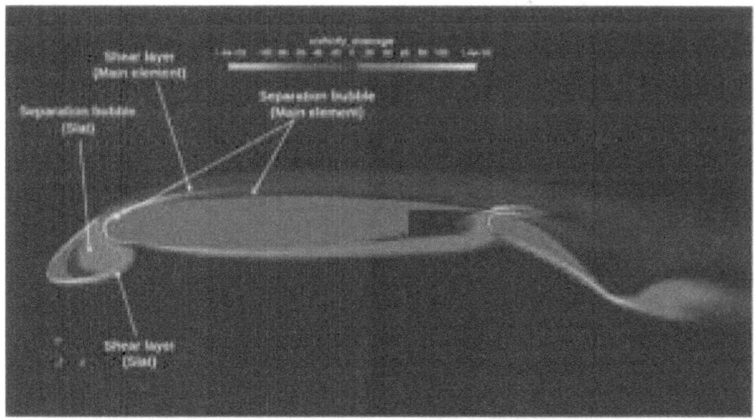

Figure 4.20: Time averaged Z-vorticity contours for flow around the 30P30N airfoil for $Re_c = 1.27 \times 10^4$ and AOA = 4° (2D case)

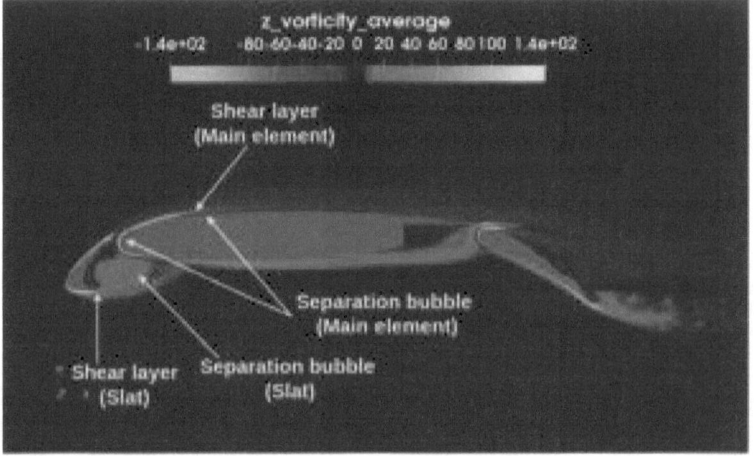

Figure 4.21: Time averaged Z-vorticity contours for flow around the 30P30N wing for $Re_c = 1.27 \times 10^4$ and AOA $= 4°$ at midspan (3D case)

4.3.3 $Re_c = 1.83 \times 10^4$

Figures 4.22, 4.23, and 4.24 represent the instantaneous vorticity contours for 2D (Figure 4.22) and 3D (Figures 4.23, 4.24) simulations for $Re_c = 1.83 \times 10^4$. Figures 4.25 and 4.26 represent time-averaged vorticity contours for 2D and 3D, respectively.

In both 2D and 3D cases, a shear layer labeled "Shear layer 1 (slat)" (Figures 4.22, 4.23, 4.24, 4.25, 4.26) forms on the leading edge of the slat. No roll-up is observed in the slat cove area from the "Shear layer (slat)" in the case of the 2D simulation, in contrast to the experimental results [3]. With the absence of roll-up, the "Shear layer (slat)" reattaches in the laminar state and forms a laminar separation bubble labeled "Separation bubble (slat)" in the slat cove (Figures 4.22, 4.25). Nonetheless, this roll-up is clearly observed in the 3D simulation that matches well with the experimental data [3] (Figures 4.23, 4.24, 4.26). This explains the limitation of the 2D simulation in this case.

Also, the presence of a boundary layer labeled "Boundary layer (main element)" (rather than a shear layer) is noticed on top of the main element due to no leading-edge separation on the main element in this case. This behaviour differs from what is observed in the case of $Re_c = 8.32 \times 10^3$ and 1.27×10^4 where a shear layer dominates on top of the main element. The shed spanwise vortices are observed coming from the "Shear layer (slat)" in case of the 3D simulation. These shed vortices deform after coming into contact with the trailing edge of the slat. Further, these shed vortices accelerate in the gap flow between the slat and main element, and persist in the slat wake. Shed spanwise vortices are also observed on top of the main element. These come from shedding of the boundary layer (main element) on top of the main element and interaction of shed spanwise vortices coming from the slat cove region with the boundary layer (main element) (Figures 4.23, 4.24, 4.26).

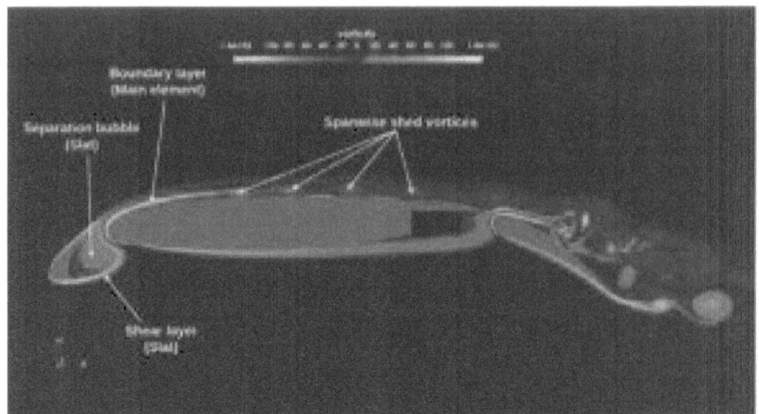

Figure 4.22: Instantaneous Z-vorticity contours for flow around the 30P30N airfoil for $Re_c = 1.83 \times 10^4$ and AOA $= 4°$ at $tU_\infty/c = 70.60$ (2D case)

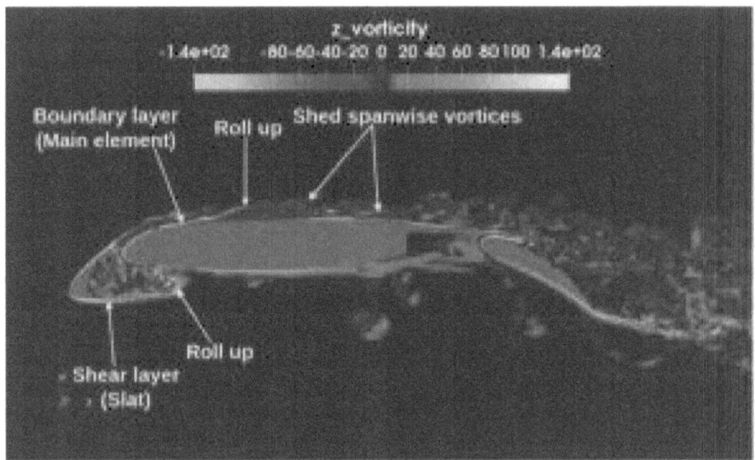

Figure 4.23: Instantaneous Z-vorticity contours for flow around the 30P30N wing for $Re_c = 1.83 \times 10^4$ and AOA $= 4°$ at midspan and $tU_\infty/c = 10.40$ (3D case)

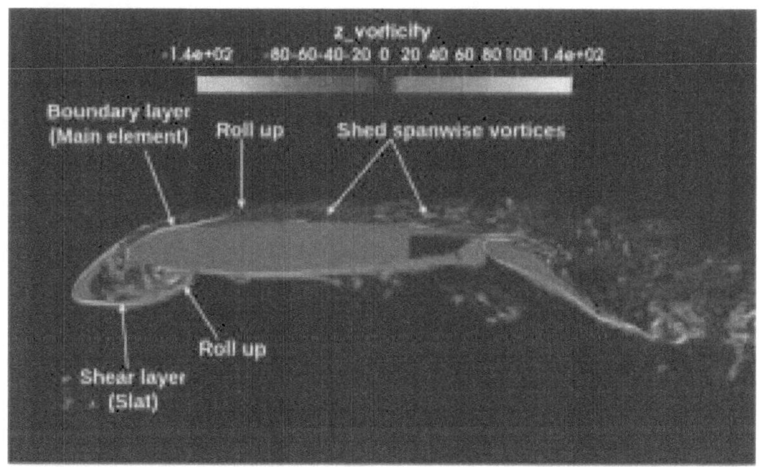

Figure 4.24: Instantaneous Z-vorticity contours for flow around the 30P30N wing for $Re_c = 1.83 \times 10^4$ and AOA = 4° at midspan and $tU_\infty/c = 13.35$ (3D case)

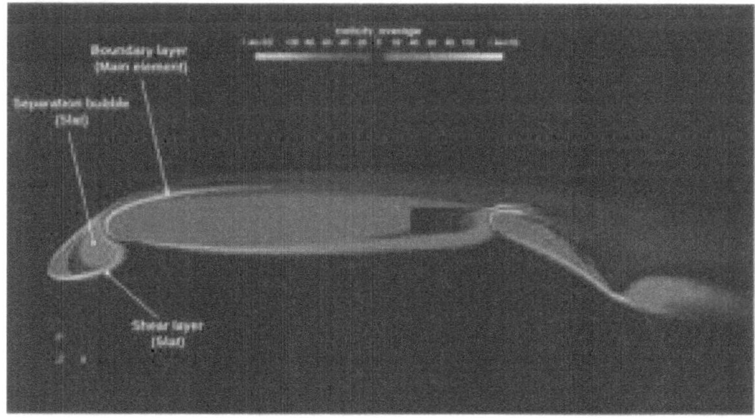

Figure 4.25: Time averaged Z-vorticity contours for flow around the 30P30N airfoil for $Re_c = 1.83 \times 10^4$ (2D case)

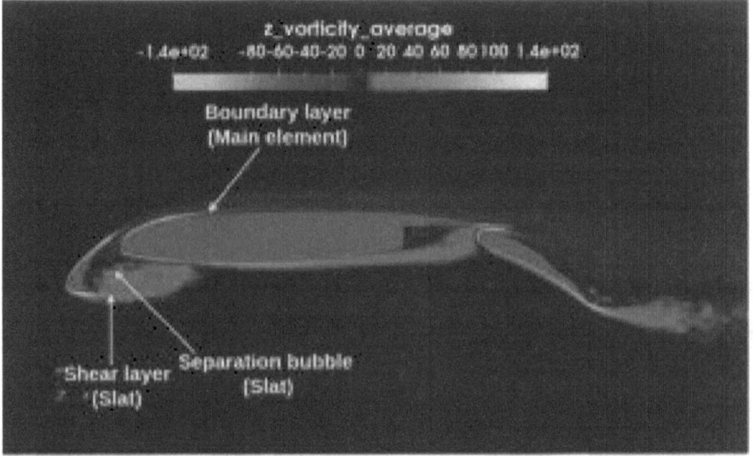

Figure 4.26: Time averaged Z-vorticity contours for flow around the 30P30N wing for $Re_c = 1.83 \times 10^4$ at midspan (3D case)

4.4 3D Vortical Flow Structures

This section discusses the three-dimensional vortical structures observed in the slat cove region when flow transition occurs from two dimensions to three dimensions.

Transition from two-dimensional to three-dimensional flow in flow past circular and square cylinders has been intensively studied in the literature e.g. [73] [74] [75] [76]. Three-dimensional instabilities have been divided into three types. First, the instability having the longest wavelength, mode A, is characterized by tongue-shaped streamwise vortical structures for Reynolds numbers (based on the diameter of the cylinder) between 160 and 175 [77] [75]. Second, the instability with the shortest wavelength, mode B, is identified by rib-like vortical structures for Reynolds numbers between 190 and 250 [77] [75]. Third, the instability with the wavelength between modes A and B, mode C, is observed at Reynolds numbers higher than those of mode A and mode B [78].

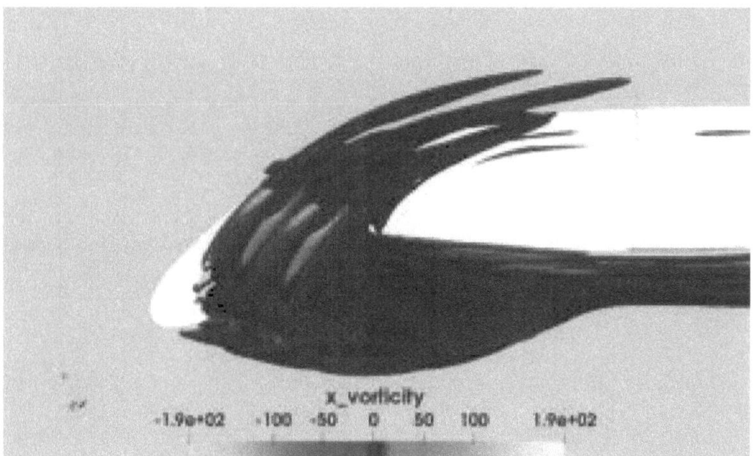

Figure 4.27: Tongue-shaped vortical structures identified in the slat cove region by instantaneous X-vorticity contours for flow around the 30P30N wing for $Re_c = 8.32 \times 10^3$ and AOA $= 4°$ at $tU_\infty/c = 14.19$ (3D case)

All simulations performed in the present study contain no forcing upstream disturbances. This means the transition from two dimensions to three dimensions is encouraged by discretization errors emerging from the governing equations or boundary conditions. This implies a natural presence for instability modes.

At $Re_c = 8.32 \times 10^4$, tongue-shaped vortical structures are observed in the spanwise direction by plotting instantaneous X-vorticity contours as shown in Figure 4.27. These structures are similar to those characterizing mode A instability observed in flow past a cylinder. The measured non-dimensional wavelength (λ/d, where d is the characteristic length) of tongue-shaped vortical structures, in this case, is 1.15. In the case of a square cylinder, characteristic length is equal to the length of a side of the cylinder that is the distance between two separation points [73] [74]. Similarly, in the present case of the 30P30N wing, the distance between two separation points on the slat was taken as characteristic length; this holds for all presented cases at different Reynolds numbers. Time evolution of tongue-shaped vortical structures is shown in Figure 4.28.

For $Re_c = 1.27 \times 10^4$ and 1.83×10^4, rib-like vortical structures are observed in the spanwise direction by plotting instantaneous X-vorticity contours as shown in Figures 4.29 and 4.30, respectively. These structures are similar to those characterizing mode B instability observed in flow past a cylinder. The measured non-dimensional wavelengths of rib-like vortical structures in these cases are 0.51 and 0.29, respectively. Figures 4.31 and 4.32 represent time evolution of rib-like vortical structures. For $Re_c = 1.83 \times 10^4$, streaks start to appear between Figures 4.32b and 4.32c on top of the main element.

The measured non-dimensional wavelength in the present case of a multi-element airfoil differs from the square cylinder case. For flow past a square cylinder, the measured non-dimensional wavelength range for mode A is 5 - 5.8 [77], and for mode B is 1.2 - 1.4 [77]. The difference in the non-dimensional wavelength range is due to the asymmetric nature of a multi-element wing compared to a square cylinder or possibly from the characteristic length chosen for the non-dimensionalization. Nevertheless, it is clear from the vorticity contours that the structures have a marked difference in wavelength value between those resembling mode A (tongue-like) and those resembling mode B (rib-like).

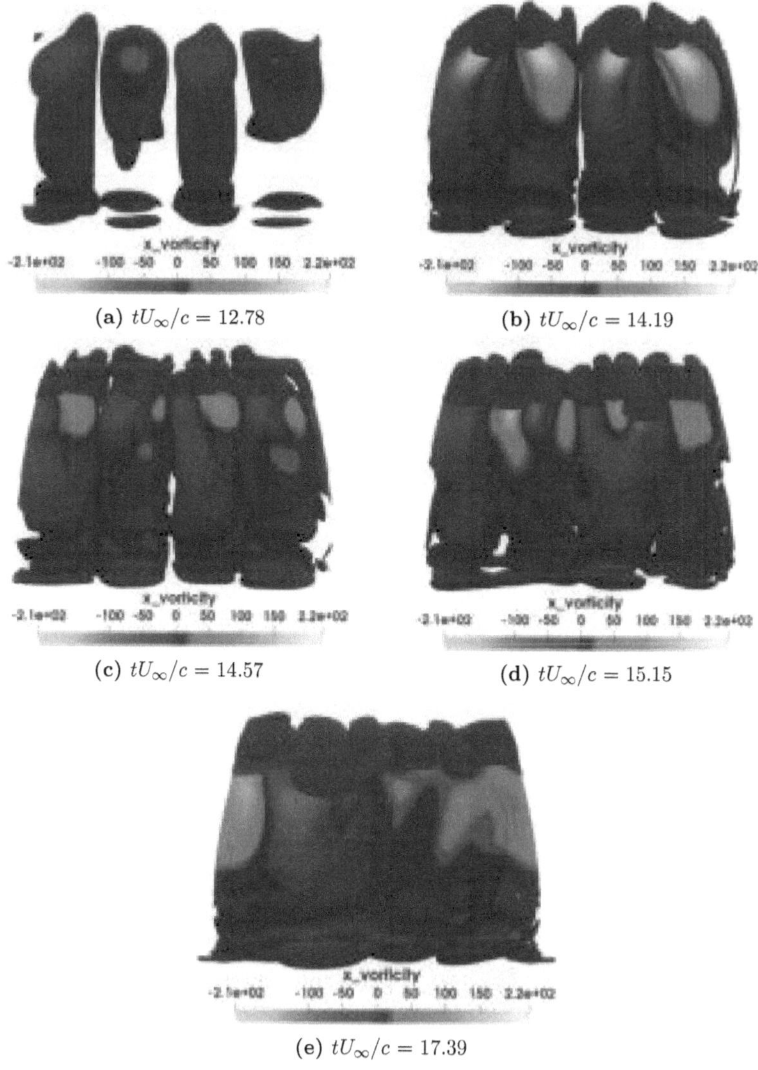

(a) $tU_\infty/c = 12.78$

(b) $tU_\infty/c = 14.19$

(c) $tU_\infty/c = 14.57$

(d) $tU_\infty/c = 15.15$

(e) $tU_\infty/c = 17.39$

Figure 4.28: Time evolution of tongue-shaped vortical structures identified in the slat cove region by front view of instantaneous X-vorticity contours for flow around the 30P30N wing for $Re_c = 8.32 \times 10^3$ and AOA $= 4°$ (3D case)

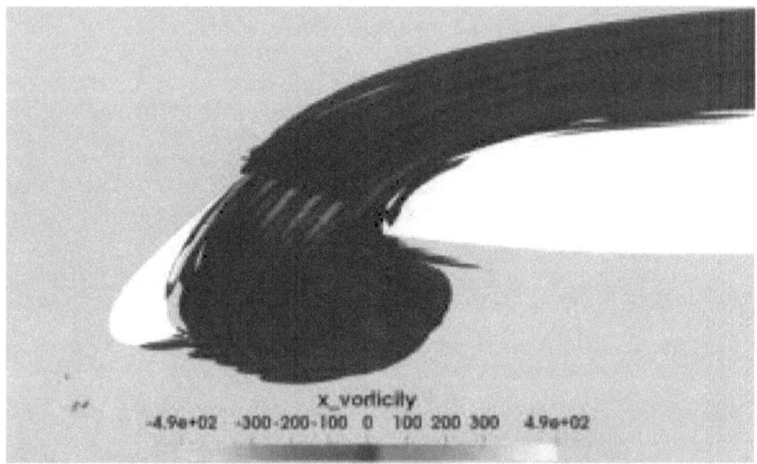

Figure 4.29: Rib-like vortical structures identified in the slat cove region by instantaneous X-vorticity contours for flow around the 30P30N wing for $Re_c = 1.27 \times 10^4$ and AOA $= 4°$ at $tU_\infty/c = 7.23$ (3D case)

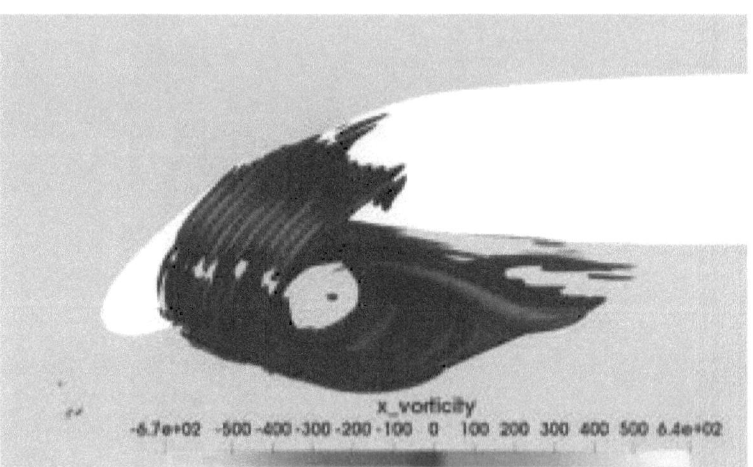

Figure 4.30: Rib-like vortical structures identified in the slat cove region by instantaneous X-vorticity contours for flow around the 30P30N wing for $Re_c = 1.83 \times 10^4$ and AOA $= 4°$ at $tU_\infty/c = 4.21$ (3D case)

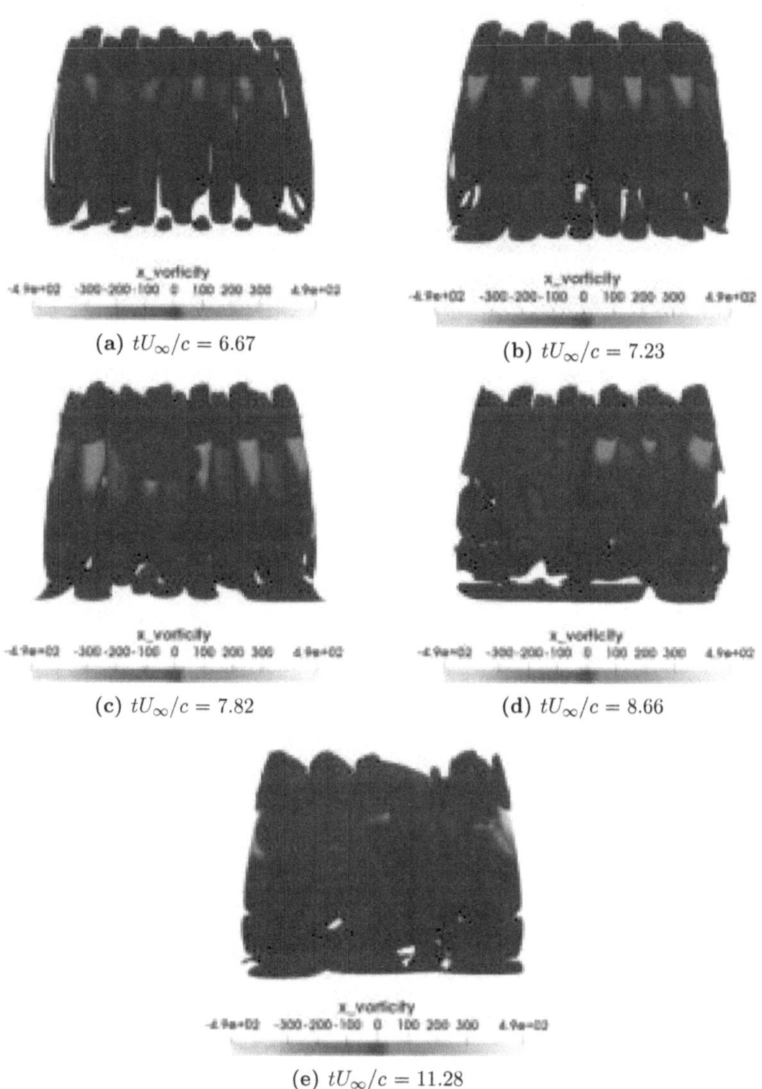

(a) $tU_\infty/c = 6.67$

(b) $tU_\infty/c = 7.23$

(c) $tU_\infty/c = 7.82$

(d) $tU_\infty/c = 8.66$

(e) $tU_\infty/c = 11.28$

Figure 4.31: Time evolution of rib-like vortical structures identified in the slat cove region by front view of instantaneous X-vorticity contours for flow around the 30P30N wing for $Re_c = 1.27 \times 10^4$ and AOA $= 4°$ (3D case)

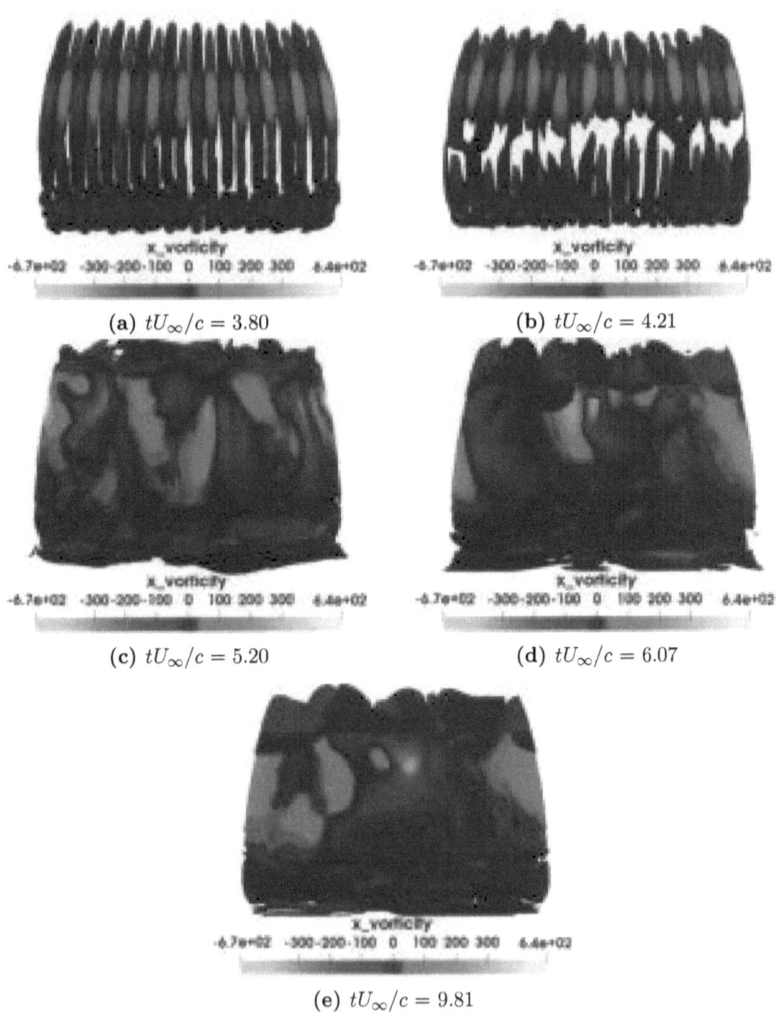

(a) $tU_\infty/c = 3.80$

(b) $tU_\infty/c = 4.21$

(c) $tU_\infty/c = 5.20$

(d) $tU_\infty/c = 6.07$

(e) $tU_\infty/c = 9.81$

Figure 4.32: Time evolution of rib-like vortical structures identified in the slat cove region by front view of instantaneous X-vorticity contours for flow around the 30P30N wing for $Re_c = 1.83 \times 10^4$ and AOA = 4° (3D case)

4.5 Görtler Vortices

Görtler vortices are pairs of counter-rotating streamwise vortices observed on top of a concave surface [79] [80] [81]. The following discussion has been restricted to the case of $Re_c = 1.27 \times 10^4$ for conciseness as similar trends are also observed in the case of $Re_c = 8.32 \times 10^3$ (supplementing figures are given in Appendix C). The presence of Görtler vortices, proposed by Wang et al. [2] [3] from the experimental results, is confirmed by plotting contours of vorticity magnitude for $Re_c = 1.27 \times 10^4$ as shown in Figure 4.33. This gives a description of the Görtler vortices in three dimensions. Usually, the presence of Görtler vortices is noted in a concave boundary layer whereas in this case it is observed on top of the convex surface (main element) [79] [80] [81]. These Görtler vortices emerge from the gap between the slat and main element and prevail on top of the main element before vortex shedding occurs. Figure 4.34 represents the Görtler vortices observed by hydrogen bubble flow visualization in experimental settings for $Re_c = 1.27 \times 10^4$ [3].

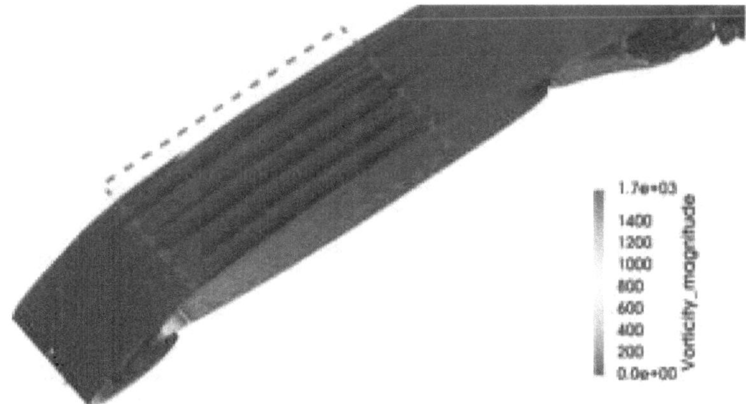

Figure 4.33: Görtler vortices identified by contours of vorticity magnitude (isometric view from the top) for flow around the 30P30N three-element wing at $Re_c = 1.27 \times 10^4$ and $tU_\infty/c = 7.20$ (3D case)

Figure 4.35 shows a sketch of pairwise counter-rotating Görtler vortices gen-

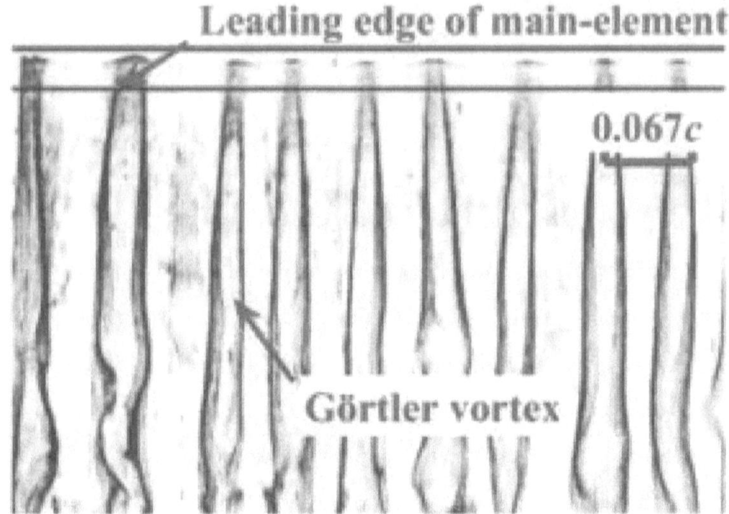

Figure 4.34: Görtler vortices identified through hydrogen bubble flow visualization for flow around the 30P30N three-element wing at $Re_c = 1.27 \times 10^4$ [3]

erated by a slightly concave surface [82]. For $Re_c = 8.32 \times 10^3$ and 1.27×10^4, a separation bubble exists in the slat cove region, as mentioned in Section 4.3, that forms a (virtual) curved wall responsible for the generation of Görtler vortices as suggested by Wang et al. [2] [3] (Figure 4.36). The spanwise motion of Görtler vortices is caused by the spanwise movement of flow inside the slat cusp separation bubble. Later, these Görtler vortices come into contact with the separated shear layer on top of the main element and result in complex interactions.

Figure 4.37 shows the top view of the Görtler vortices. For this case the measured non-dimensional wavelength (λ_g/c) of Görtler vortices [81] is 0.029 before the spanwise deviation occurs in the position of Görtler vortices which is of the same order of magnitude as shown in the experimental images of Wang et al. [3].

The conventional "Mushroom shape" of Görtler vortices [83] is identified by plotting cross sections of vorticity magnitude over multiple locations in the stream wise direction as shown in Figure 4.38. Also, this "Mushroom shape" of Görtler

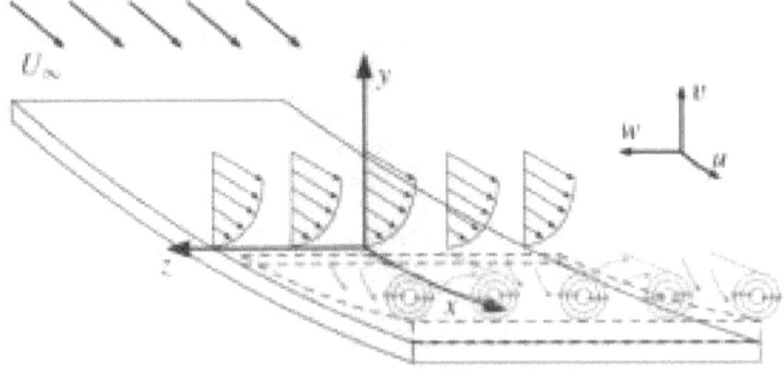

Figure 4.35: Sketch of the counter-rotating Görtler vortices observed on a concave surface [82]

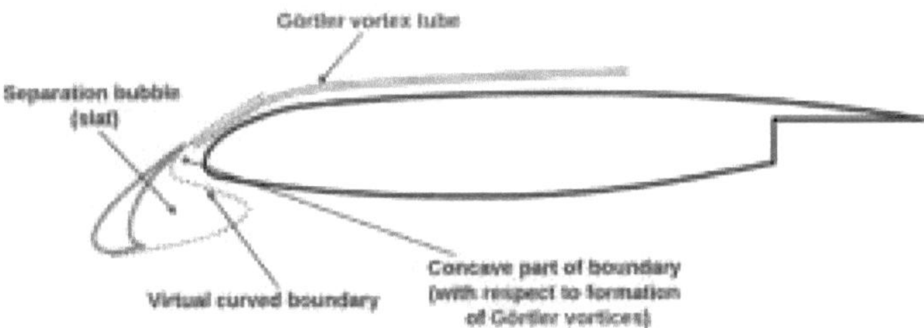

Figure 4.36: Sketch of the Görtler vortex tube generated due to presence of a virtual curved boundary in the slat cove region

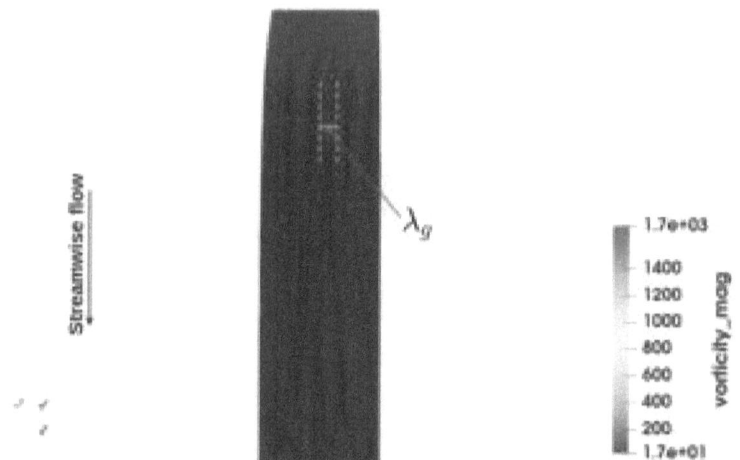

Figure 4.37: Görtler vortices identified by contours of vorticity magnitude (top view) for flow around the 30P30N three-element wing at $Re_c = 1.27 \times 10^4$ and $tU_\infty/c = 9.20$ (3D case). λ_g indicates the (spanwise) wavelength of the Görtler vortices

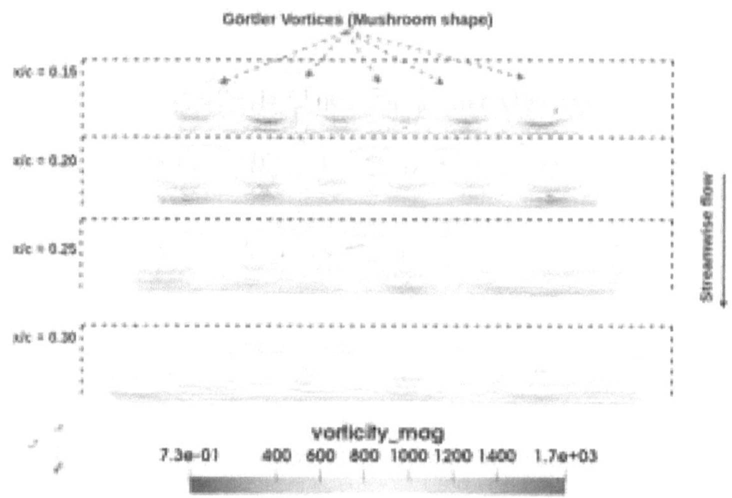

Figure 4.38: Conventional "Mushroom shape" of Görtler vortices identified by cross-sectional contours of vorticity magnitude at different locations in the streamwise direction for flow around the 30P30N three-element wing at $Re_c = 1.27 \times 10^4$ and $tU_\infty/c = 9.20$ (3D case)

vortices persists along the flow in streamwise direction.

Pair-wise counter-rotating motion of Görtler vortices is shown in Figure 4.39 with the help of green arrows. This counter-rotating motion is observed in both cases of concave [79] [80] [81] and convex walls [2]. Secondary counter-rotating vortices are spotted underneath the primary Görtler vortices and shown with the help of yellow arrows in Figure 4.39. The Görtler vortices above the main element induce these secondary counter-rotating vortices due to the presence of the wall boundary (due to viscous effect). Compared to the primary Görtler vortices, these secondary counter-rotating vortices have opposite rotation directions. These secondary counter-rotating vortices lie under the separated shear layer, whereas primary Görtler vortices reside on top of the separated shear layer [2].

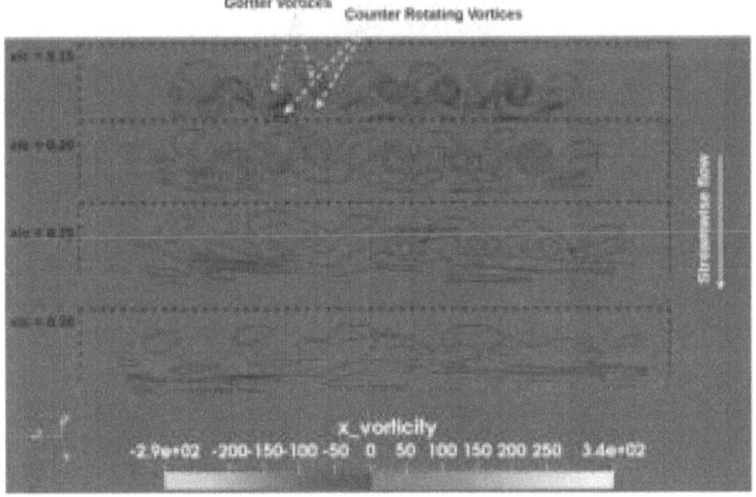

Figure 4.39: Counter-rotating vortices identified beneath Görtler vortices by cross-sectional contours of X-vorticity at different locations in streamwise direction for flow around the 30P30N three-element wing at $Re_c = 1.27 \times 10^4$ and $tU_\infty/c = 9.20$ (3D case)

Figures 4.40 and 4.41 identify various patterns of spanwise motion of the Görtler vortices. This unpredictable spanwise motion observed in Görtler vortices is very distinct from the conventional static Görtler vortices case [83] [84].

In Figure 4.40, pattern A indicates that one pair of counter-rotating Görtler vortices can split and result into two pairs of counter-rotating Görtler vortices. On the other hand, pattern B indicates the spanwise rightward motion of Görtler vortices. In Figure 4.41, pattern D indicates that two pairs of counter-rotating Görtler vortices can merge and result into one pair of counter-rotating Görtler vortices, whereas pattern C indicates the spanwise leftward motion of Görtler vortices.

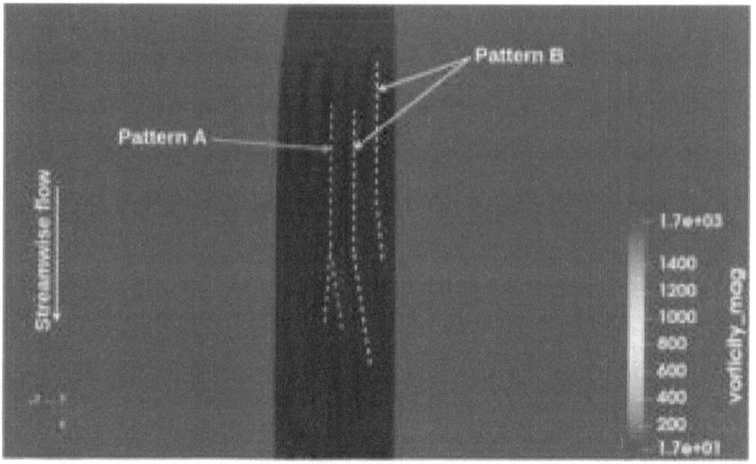

Figure 4.40: Spanwise motion of counter-rotating Görtler vortices identified by contours of vorticity magnitude for flow around the 30P30N three-element wing at $Re_c = 1.27 \times 10^4$ and $tU_\infty/c = 9.20$ (3D case): pattern A indicates the splitting and pattern B indicates the rightward motion of Görtler vortices

Figures 4.42 to 4.45 detail the evolution of Görtler vortices for $Re_c = 1.27 \times 10^4$ using instantaneous Z-vorticity contours. Initially, the flow is nearly two-dimensional as shown in Figure 4.42 as no three-dimensional structures are observed. Later in time, as indicated in Figure 4.43, counter-rotating Görtler vortices start to appear and secondary counter-rotating vortices are also noted beneath the original Görtler vortices. Subsequently, as detailed in Figure 4.44, both Görtler vortices and secondary counter-rotating vortices evolve and become prominent. Finally, Figure 4.45 indicates the presence of spanwise shed vortices on top of the main element resulting from destabilization of Görtler vortices and the shear layer

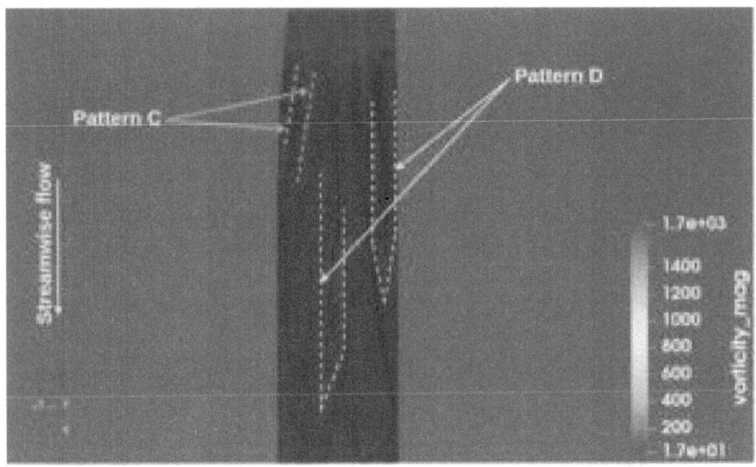

Figure 4.41: Spanwise motion of counter rotating Görtler vortices identified by contours of vorticity magnitude for flow around the 30P30N three-element wing at $Re_c = 1.27 \times 10^4$ and $tU_\infty/c = 11.30$ (3D case): pattern C indicates the leftward motion and pattern D indicates the merging of Görtler vortices

(Figures 4.18, 4.19) on top of the main element.

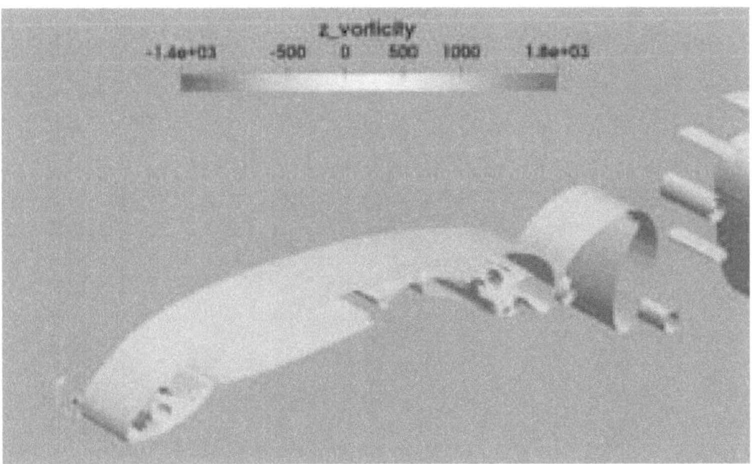

Figure 4.42: Instantaneous Z-vorticity contours for flow around the 30P30N wing for $Re_c = 1.27 \times 10^4$ at $tU_\infty/c = 5.33$ (3D case)

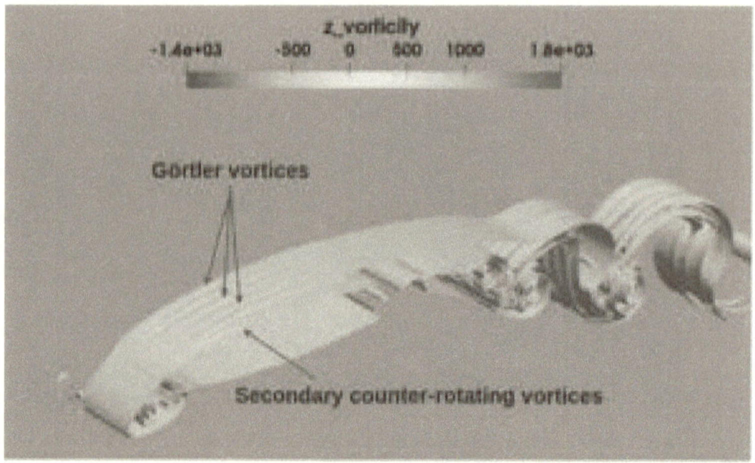

Figure 4.43: Instantaneous Z-vorticity contours for flow around the 30P30N wing for $Re_c = 1.27 \times 10^4$ at $tU_\infty/c = 6.67$ (3D case)

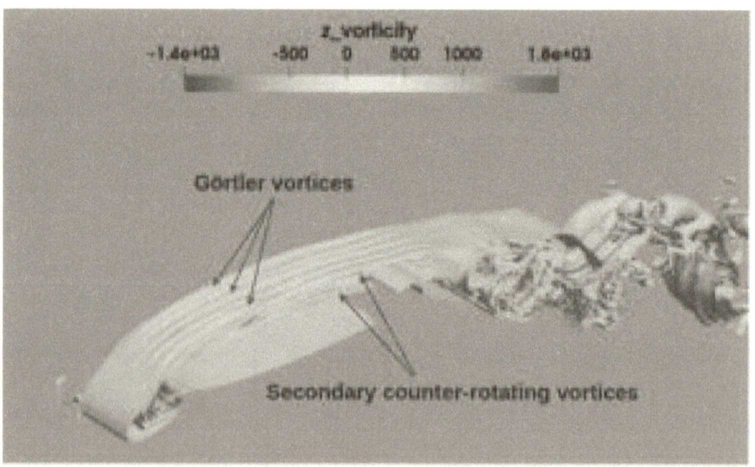

Figure 4.44: Instantaneous Z-vorticity contours for flow around the 30P30N wing for $Re_c = 1.27 \times 10^4$ at $tU_\infty/c = 7.82$ (3D case)

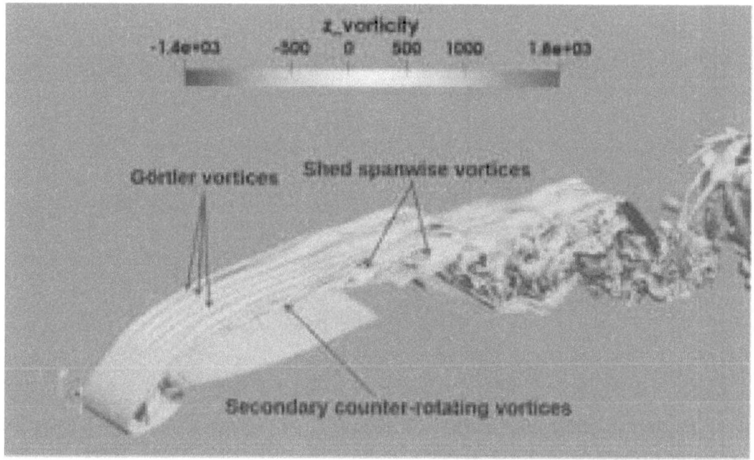

Figure 4.45: Instantaneous Z-vorticity contours for flow around the 30P30N wing for $Re_c = 1.27 \times 10^4$ at $tU_\infty/c = 8.84$ (3D case)

4.6 Hairpin Vortices

The presence of hairpin vortices was confirmed by plotting contours of vorticity magnitude for $Re_c = 1.83 \times 10^4$ as shown in Figure 4.46. This gives a description of the hairpin vortices in three dimensions. Figure 4.47 represents the hairpin vortices observed by hydrogen bubble flow visualization in experimental settings for the same Reynolds number, $Re_c = 1.83 \times 10^4$ [3].

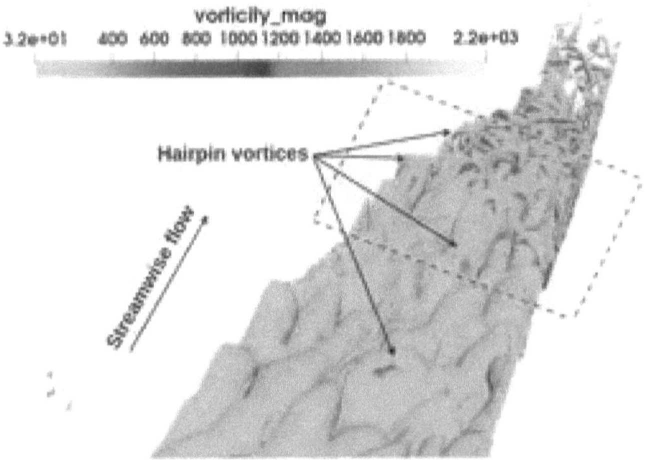

Figure 4.46: Hairpin vortices identified by contours of vorticity magnitude (looking downstream from above) for flow around the 30P30N three-element wing at $Re_c = 1.83 \times 10^4$ and $tU_\infty/c = 13.20$ (3D case)

For $Re_c = 1.83 \times 10^4$, shed spanwise vortices from the slat cove dominate in the gap region and are later deformed by the impact on the slat trailing edge and the accelerated flow in the gap region (Figures 4.23, 4.24). These deformed vortices later evolve into hairpin-like vortices as shown in Figure 4.46. As indicated in Figure 4.48, heads of the hairpin vortices act like spanwise vortices, whereas tails of the hairpin vortices act like streamwise vortices. This confirms the existence of both streamwise and spanwise vortices in the wake [3].

Figure 4.49 shows Acarlar and Smith's sketch explaining hairpin vortex formation in the laminar boundary layer [85]. They generated synthetic low-speed

Figure 4.47: Hairpin vortices identified through hydrogen bubble flow visualization (top view) for flow around the 30P30N three-element wing at $Re_c = 1.83 \times 10^4$ [3]

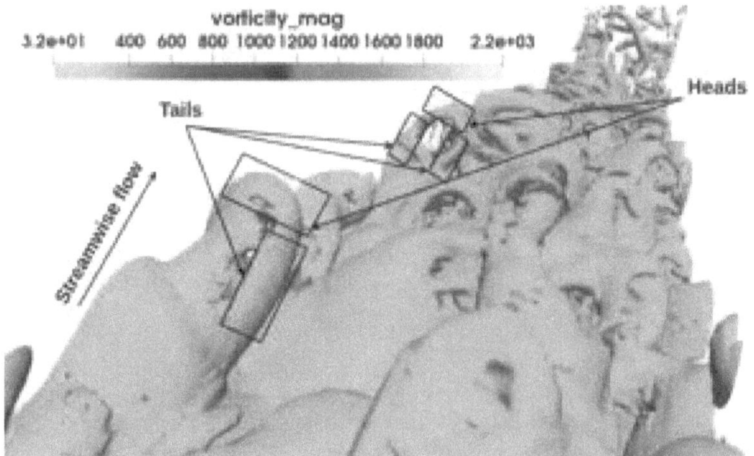

Figure 4.48: Close-up of hairpin vortices identified by contours of vorticity magnitude (looking downstream from above) for flow around the 30P30N three-element wing at $Re_c = 1.83 \times 10^4$ and $tU_\infty/c = 13.20$ (3D case)

streaks by fluid injection in the laminar boundary layer. As the injection speed is increased above critical values, horse-shoe type vortices start appearing and are carried along with the flow. These vortices undergo a stretching process and result in hairpin vortices.

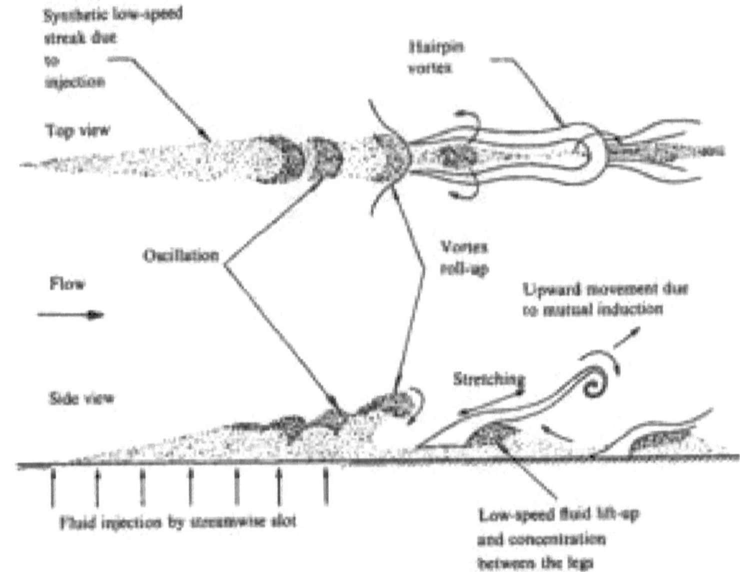

Figure 4.49: Sketch of the hairpin vortex formation observed in a laminar boundary layer [85]

A similar phenomenon is observed in the present case of flow over the 30P30N wing at $Re_c = 1.83 \times 10^4$. This is described by instantaneous Z-vorticity contours in Figures 4.50 to 4.53. Initially, the flow is two-dimensional, as shown in Figure 4.50 as no three-dimensional coherent structures are discernible. Later, as indicated in Figure 4.51, streaky structures start to appear on top of the main element. These could be due to the interaction between the shear layer coming from the top of the slat, spanwise vortices generated in the slat cove, acceleration through the gap region between the slat and the leading edge of the main element, and finally the boundary layer on the top of the main element. The spanwise vortices might play a role as an external flow injection similar to the synthetic ones of Acarlar and Smith shown in Figure 4.49. Subsequently, vortex roll-up is observed in this streaky structure downstream on the main element. Finally, as shown in Figure 4.53, this results in the formation of hairpin vortices. These generated hairpin vortices go through a regenerative process and can spawn new hairpin vortices [86].

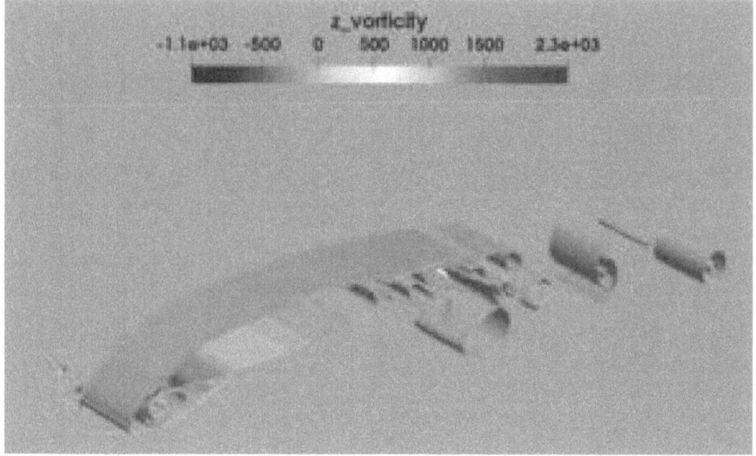

Figure 4.50: Instantaneous Z-vorticity contours for flow around the 30P30N wing for $Re_c = 1.83 \times 10^4$ at $tU_\infty/c = 4.21$ (3D case)

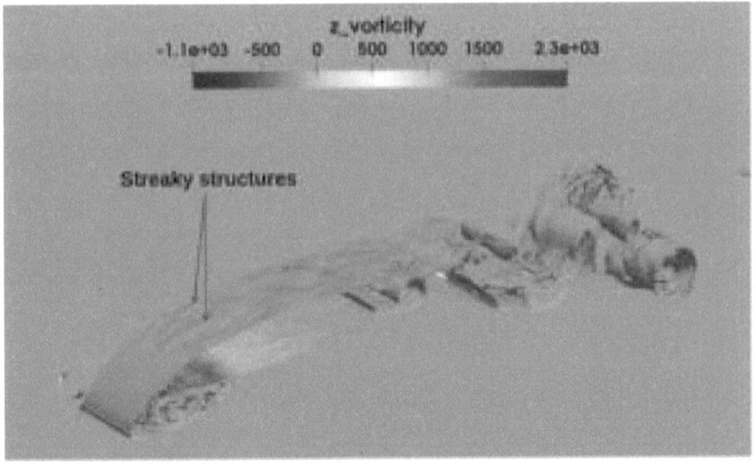

Figure 4.51: Instantaneous Z-vorticity contours for flow around the 30P30N wing for $Re_c = 1.83 \times 10^4$ at $tU_\infty/c = 5.37$ (3D case)

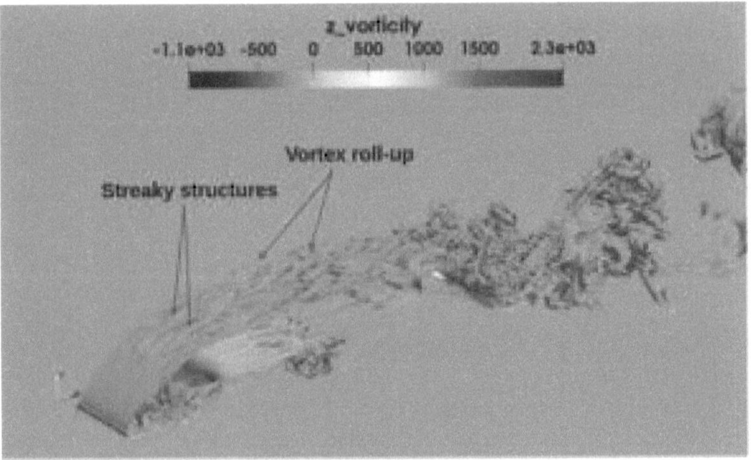

Figure 4.52: Instantaneous Z-vorticity contours for flow around the 30P30N wing for $Re_c = 1.83 \times 10^4$ at $tU_\infty/c = 6.21$ (3D case)

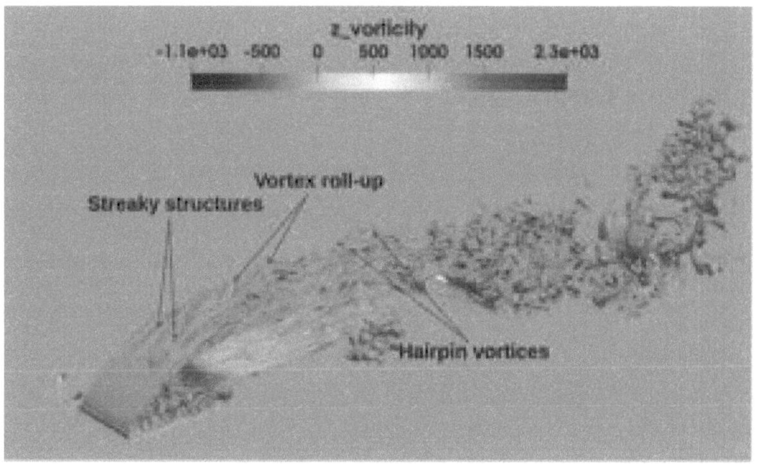

Figure 4.53: Instantaneous Z-vorticity contours for flow around the 30P30N wing for $Re_c - 1.83 \times 10^4$ at $tU_\infty/c = 6.60$ (3D case)

Chapter 5

Conclusion and Future Work

5.1 Conclusion

The flow physics behind multi-element airfoils and wings are critical to understand in order to further improve performance of aircraft, both large and small. To get an accurate prediction of flow, high-order methods are needed to perform direct numerical simulations. The complex geometry of the 30P30N three-element high lift system poses a challenge for high-order spectral element simulations. For example, in the present case of the Nek5000 spectral element code, structured grid generation was a crucial and challenging task. Further, the sheer size of the data sets produced presents a significant challenge to developing useful insights about the flow. Successful simulation of flow in this intricate geometry presents a major achievement for high-order methods. Moreover, the data generated by high-order simulations can serve as an excellent data set for validation of other numerical methods such as RANS, LES, and DES, etc.

2D and 3D simulations have been performed successfully for $Re_c = 8.32 \times 10^3$, 1.27×10^4, and 1.83×10^4 flows over the 30P30N high lift system. In the case of $Re_c = 8.32 \times 10^3$, the shear layers coming from both top and bottom surfaces of the slat, a separation bubble in the slat cove region and on top of the main element, and a shear layer on the main element due to the leading edge separation have been successfully predicted. Roll-up and corresponding spanwise shed vortices have been observed on top of the main element. Similarly, for $Re_c = 1.27 \times 10^4$, a shear layer has been observed to come from the bottom of the slat and top of the main element. Two separation bubbles have been identified, one in the slat cove and second on top of the main element. The roll-up and subsequent shed vortices were observed in this case as well. However, for $Re_c = 1.83 \times 10^4$, no separation bubble was found on top of the main element, which indicates the presence of a boundary layer (rather than a shear layer) on top of the main element. A roll-up was observed in the slat cove that led to the generation of spanwise vortices in slat cove, and these persisted in the slat wake. All these results were not reproduced exactly in 2D simulations; this implies that 2D simulations are not sufficient to capture the flow physics in this case.

Time-averaged drag and lift coefficients were calculated, showing an increase in lift with increasing Reynolds number, as expected, and a decrease in drag coefficient with Reynolds number in the critical Reynolds number range studied. Detailed pressure coefficient plots support these results.

Different three-dimensional vortical structures were identified in the transition of flow from two dimensions to three dimensions. Tongue-shaped vortical structures (similar to mode A) were observed for $Re_c = 8.32 \times 10^3$, whereas rib-like vortical structures (similar to mode B) were present for both $Re_c = 1.27 \times 10^4$ and 1.83×10^4. These vortical structures were also observed in flow past circular and square cylinders [73] [74] [75] [76].

The presence of Görtler vortices was observed on top of the convex surface (main element) for $Re_c = 8.32 \times 10^3$ and 1.27×10^4. Secondary counter-rotating vortices were also observed beneath the main Görtler vortices just above the main element surface. The Görtler vortices were observed to travel in the spanwise direction, as well as to merge and split, which is quite different from the typical Görtler vortices studies where they are formed by flow along a concave surface [79] [80] [81]. A (virtual) curved wall formed by the separation bubble in the slat cove region was thought to be responsible for the generation of Görtler vortices [2] [3]. The formation of these Görtler vortices was also illustrated through time-varying Z-vorticity contours. Secondary vortices were also observed, which also move in the spanwise direction.

Hairpin vortices were identified on top of the main element for $Re_c = 1.83 \times 10^4$. This confirms the presence of both streamwise and spanwise vortices on top of the main element. The generation of these hairpin vortices was illustrated through time-varying Z-vorticity contours that resemble the formation of hairpin vortices in the laminar boundary layer [87] [85].

All the presented results are in very good agreement with the experimental data [2] [3] and are a first numerical simulation of the complex three-dimensional flow in high lift systems at moderate Reynolds numbers. These are the only results on this geometry that do not use a turbulence model. While the Reynolds numbers chosen do not yet exhibit fully turbulent flow, this work represents a major step towards direct numerical simulation of high lift systems at operational Reynolds numbers of commercial aircraft. The Reynolds numbers studied are more representative of small aerial vehicles and, as such, these results detail those flows and will allow for better understanding of performance of small aerial vehicles. In addition to reproducing experimental results, new contributions were made in the following areas: 1) lift and drag analysis was provided; 2) pressure coefficient curves for each element of the multi-element airfoil were provided; 3) three-dimensional vortical structures were identified and detail the transition from 2D to 3D flow; 4) a more detailed time evolution of the Görtler vortices was provided, beyond what was observed in the experiments, adding downstream splitting and merging of these vortices; and 5) time evolution of hairpin vortices was more clearly visualized than in the experiments through 3D rendering of vorticity surfaces and their formation was explained through the development of streaky structures. All of these results contribute to a better understanding of the flow physics for high lift devices operating at low Reynolds numbers and provide insight for the formation of turbulent higher Reynolds number flows where commercial aircraft operate.

5.2 Future Work

While the major mechanisms of the complex flow past a 30P30N high-lift system have been identified, the data sets obtained are rich and provide opportunities for further exploration. For example, the flap wake and the physics associated with it have not been studied in this work, yet they will be very useful in understanding lift and drag characteristics as well as noise production which is a significant concern for high-lift. The 3D simulations were very long and expensive to run and, as such, were not run long enough to provide sufficient statistics for frequency analysis and turbulent characteristics analysis. Further work should be planned around running these simulations for longer times to allow for these analyses. The ultimate goal is to run simulations for higher Reynolds numbers. The approach for the grids developed in this work is estimated to be suitable for $Re_c = 9.0 \times 10^6$ as the value of y^+ turbulent has been found to be less than one in RANS calculations performed using the same multi-block approach. The computation speed can also be increased by using the recently released GPU-compatible version of Nek5000 called NekRS [88].

References

[1] P. Fischer, J. Lottes, S. Kerkemeier, A. Obabko, and K. Heisey, "Nek5000 [Software]. Mathematics and Computer Science Division of Argonne National Laboratory." https://nek5000.mcs.anl.gov/. (Retrieved on 04/10/2019).

[2] J. S. Wang, L. H. Feng, J. J. Wang, and T. Li, "Görtler vortices in low-Reynolds-number flow over multi-element airfoil," *Journal of Fluid Mechanics*, vol. 835, pp. 898–935, 2018.

[3] J. Wang, J. Wang, and K. C. Kim, "Wake/shear layer interaction for low-Reynolds-number flow over multi-element airfoil," *Experiments in Fluids*, vol. 60, no. 1, p. 16, 2019.

[4] P. K. Rudolph, "High-lift systems on commercial subsonic airliners," NASA-CR-4746, 1996.

[5] D. Butter, "Recent progress on development and understanding of high lift systems," AGARD-CP-365, Paper 1, 1984.

[6] P. Garner, P. Meredith, and R. Stoner, "Areas for future CFD development as illustrated by transport aircraft applications," in *10th Computational Fluid Dynamics Conference*, 1991. doi:10.2514/6.1991-1527.

[7] C. Van Dam, "The aerodynamic design of multi-element high-lift systems for transport airplanes," *Progress in Aerospace Sciences*, vol. 38, no. 2, pp. 101–144, 2002.

[8] C. L. Rumsey and S. X. Ying, "Prediction of high lift: review of present CFD capability," *Progress in Aerospace Sciences*, vol. 38, no. 2, pp. 145–180, 2002.

[9] V. Chin, D. Peters, F. Spaid, and R. Mcghee, "Flowfield measurements about a multi-element airfoil at high Reynolds numbers," in *23rd Fluid Dynamics, Plasmadynamics, and Lasers Conference*, 1993. doi:10.2514/6.1993-3137.

[10] S. Klausmeyer and J. Lin, "An experimental investigation of skin friction on a multi-element airfoil," in *12th Applied Aerodynamics Conference*, 1994. doi:10.2514/6.1994-1870.

[11] A. Bertelrud, "Transition on a three-element high lift configuration at high Reynolds numbers," in *36th AIAA Aerospace Sciences Meeting and Exhibit*, 1998. doi:10.2514/6.1998-703.

[12] C. L. Rumsey, E. M. Lee-Rausch, and R. D. Watson, "Three-dimensional effects in multi-element high lift computations," *Computers & Fluids*, vol. 32, no. 5, pp. 631–657, 2003.

[13] F. Lynch, R. Potter, and F. Spaid, "Requirements for effective high lift CFD," in *20th ICAS Congress*, pp. 1479–1482, 1996.

[14] S. M. Klausmeyer and J. C. Lin, "Comparative results from a CFD challenge over a 2D three-element high-lift airfoil," NASA Technical Memorandum 112858, 1997.

[15] S. Rogers, F. Menter, P. Durbin, and N. Mansour, "A comparison of turbulence models in computing multi-element airfoil flows," in *32nd Aerospace Sciences Meeting and Exhibit*, 1994. doi:10.2514/6.1994-291.

[16] W. O. Valarezo and D. J. Mavriplis, "Navier-Stokes applications to high-lift airfoil analysis," *Journal of Aircraft*, vol. 32, no. 3, pp. 618–624, 1995.

[17] C. J. Dominik, "Application of the incompressible Navier-Stokes equations

to high-lift flows," in *12th AIAA Applied Aerodynamics Conference, AIAA Paper 94-1872*, 1994.

[18] K. Jones, R. Biedron, and M. Whitlock, "Application of a Navier-Stokes solver to the analysis of multi-element airfoils and wings using multi-zonal grid techniques," in *13th AIAA Applied Aerodynamics Conference*, 1995. doi:10.2514/6.1995-1855.

[19] C. L. Rumsey and P. R. Spalart, "Turbulence model behavior in low Reynolds number regions of aerodynamic flowfields," *AIAA Journal*, vol. 47, no. 4, pp. 982–993, 2009.

[20] J. L. Thomas and M. Salas, "Far-field boundary conditions for transonic lifting solutions to the Euler equations," *AIAA Journal*, vol. 24, no. 7, pp. 1074–1080, 1986.

[21] J. Bodart, J. Larsson, and P. Moin, "Large eddy simulation of high-lift devices," in *21st AIAA Computational Fluid Dynamics Conference*, 2013. doi:10.2514/6.2013-2724.

[22] R. Langtry and F. Menter, "Transition modeling for general CFD applications in aeronautics," in *43rd AIAA Aerospace Sciences Meeting and Exhibit*, 2005. doi:10.2514/6.2005-522.

[23] T. J. Mueller and J. D. DeLaurier, "Aerodynamics of small vehicles," *Annual Review of Fluid Mechanics*, vol. 35, no. 1, pp. 89–111, 2003.

[24] A. Haines, "Scale effects on aircraft and weapon aerodynamics," Advisory Group for Aerospace Research and Development (AGARD) Technical Report AGARD-AG-323, 1994.

[25] P. Lissaman, "Low-Reynolds-number airfoils," *Annual Review of Fluid Mechanics*, vol. 15, no. 1, pp. 223–239, 1983.

[26] M. S. Boutilier and S. Yarusevych, "Separated shear layer transition over an airfoil at a low Reynolds number," *Physics of Fluids*, vol. 24, no. 8, p. 084105, 2012.

[27] B. Carmichael, "Low Reynolds number airfoil survey," NASA-CR-165803, 1981.

[28] M. Gaster, "The structure and behaviour of laminar separation bubbles," Reports and Memoranda No. 3595, Aeronautical Research Council, London, England, 1967.

[29] O. Marxen and D. S. Henningson, "The effect of small-amplitude convective disturbances on the size and bursting of a laminar separation bubble," *Journal of Fluid Mechanics*, vol. 671, pp. 1–33, 2011.

[30] S. Burgmann and W. Schröder, "Investigation of the vortex induced unsteadiness of a separation bubble via time-resolved and scanning PIV measurements," *Experiments in Fluids*, vol. 45, no. 4, p. 675, 2008.

[31] R. Hain, C. Kähler, and R. Radespiel, "Dynamics of laminar separation bubbles at low-Reynolds-number aerofoils," *Journal of Fluid Mechanics*, vol. 630, pp. 129–153, 2009.

[32] L. Jones, R. Sandberg, and N. Sandham, "Direct numerical simulations of forced and unforced separation bubbles on an airfoil at incidence," *Journal of Fluid Mechanics*, vol. 602, pp. 175–207, 2008.

[33] M. Lang, U. Rist, and S. Wagner, "Investigations on controlled transition development in a laminar separation bubble by means of LDA and PIV," *Experiments in Fluids*, vol. 36, no. 1, pp. 43–52, 2004.

[34] O. Marxen, M. Lang, and U. Rist, "Vortex formation and vortex breakup in a laminar separation bubble," *Journal of Fluid Mechanics*, vol. 728, pp. 58–90, 2013.

[35] B. R. McAuliffe and M. I. Yaras, "Transition mechanisms in separation bubbles under low-and elevated-freestream turbulence," *Journal of Turbomachinery*, vol. 132, no. 1, 2010.

[36] D. Simoni, M. Ubaldi, P. Zunino, D. Lengani, and F. Bertini, "An experimental investigation of the separated-flow transition under high-lift turbine blade pressure gradients," *Flow, Turbulence and Combustion*, vol. 88, no. 1-2, pp. 45–62, 2012.

[37] D. Simoni, D. Lengani, M. Ubaldi, P. Zunino, and M. Dellacasagrande, "Inspection of the dynamic properties of laminar separation bubbles: free-stream turbulence intensity effects for different Reynolds numbers," *Experiments in Fluids*, vol. 58, no. 6, p. 66, 2017.

[38] H. Hansen, P. Thiede, F. Moens, R. Rudnik, and J. Quest, "Overview about the European high lift research programme Eurolift," in *42nd AIAA Aerospace Sciences Meeting and Exhibit*, 2004. doi:10.2514/6.2004-767.

[39] W. Dobrzynski, "Almost 40 years of airframe noise research: what did we achieve?," *Journal of Aircraft*, vol. 47, no. 2, pp. 353–367, 2010.

[40] N. Ashton, A. West, and F. Mendonça, "Flow dynamics past a 30P30N three-element airfoil using improved delayed detached-eddy simulation," *AIAA Journal*, vol. 54, no. 11, pp. 3657–3667, 2016.

[41] M. M. Choudhari and M. R. Khorrami, "Effect of three-dimensional shear-layer structures on slat cove unsteadiness," *AIAA Journal*, vol. 45, no. 9, pp. 2174–2186, 2007.

[42] S. Deck and R. Laraufie, "Numerical investigation of the flow dynamics past a three-element aerofoil," *Journal of Fluid Mechanics*, vol. 732, pp. 401–444, 2013.

[43] L. Jenkins, M. Khorrami, and M. Choudhari, "Characterization of unsteady flow structures near leading-edge slat: Part 1: PIV measurements," in *10th AIAA/CEAS Aeroacoustics Conference*, 2004. doi:10.2514/6.2004-2801.

[44] C. C. Pagani Jr, D. S. Souza, and M. A. Medeiros, "Slat noise: aeroacoustic beamforming in closed-section wind tunnel with numerical comparison," *AIAA Journal*, vol. 54, no. 7, pp. 2100–2115, 2016.

[45] K. Paschal, L. Jenkins, and C. Yao, "Unsteady slat wake characteristics of a 2-D high-lift configuration," in *38th Aerospace Sciences Meeting and Exhibit*, 2000. doi:10.2514/6.2000-139.

[46] D. S. Souza, D. Rodríguez, L. G. C. Simões, and M. A. F. d. Medeiros, "Effect of an excrescence in the slat cove: Flow-field, acoustic radiation and coherent structures," *Aerospace Science and Technology*, vol. 44, pp. 108–115, 2015.

[47] L. Squire, "Interactions between wakes and boundary-layers," *Progress in Aerospace Sciences*, vol. 26, no. 3, pp. 261–288, 1989.

[48] S. Makiya, A. Inasawa, and M. Asai, "Vortex shedding and noise radiation from a slat trailing edge," *AIAA Journal*, vol. 48, no. 2, pp. 502–509, 2010.

[49] N. Kyriakides, E. Kastrinakis, S. Nychas, and A. Goulas, "Aspects of flow structure during a cylinder wake-induced laminar/turbulent transition," *AIAA Journal*, vol. 37, no. 10, pp. 1197–1205, 1999.

[50] V. Ovchinnikov, U. Piomelli, and M. M. Choudhari, "Numerical simulations of boundary-layer transition induced by a cylinder wake," *Journal of Fluid Mechanics*, vol. 547, pp. 413–441, 2006.

[51] C. Pan, J. J. Wang, P. F. Zhang, and L. H. Feng, "Coherent structures in bypass transition induced by a cylinder wake," *Journal of Fluid Mechanics*, vol. 603, pp. 367–389, 2008.

[52] A. T. Patera, "A spectral element method for fluid dynamics: Laminar flow in a channel expansion," *Journal of Computational Physics*, vol. 54, no. 3, pp. 468–488, 1984.

[53] E. M. Rønquist, *Optimal spectral element methods for the unsteady three-dimensional incompressible Navier-Stokes equations.* PhD book, Mas-sachusetts Institute of Technology, 1988.

[54] M. O. Deville, P. F. Fischer, and E. Mund, *High-order methods for incompressible fluid flow*, vol. 9. Cambridge University Press, 2002.

[55] G. E. Karniadakis, M. Israeli, and S. A. Orszag, "High-order splitting methods for the incompressible Navier-Stokes equations," *Journal of Computational Physics*, vol. 97, no. 2, pp. 414–443, 1991.

[56] M. Khorrami, M. Choudhari, and L. Jenkins, "Characterization of unsteady flow structures near leading-edge slat: Part 2: 2D computations," in *10th AIAA/CEAS Aeroacoustics Conference*, 2004. doi:10.2514/6.2004-2802.

[57] S. M. Hosseini, R. Vinuesa, P. Schlatter, A. Hanifi, and D. S. Henningson, "Direct numerical simulation of the flow around a wing section at moderate Reynolds number," *International Journal of Heat and Fluid Flow*, vol. 61, pp. 117–128, 2016.

[58] C. Geuzaine and J. Remacle, "Gmsh: a three-dimensional finite element mesh generator with built-in pre- and post-processing facilities," *International Journal for Numerical Methods in Engineering*, vol. 79, no. 11, pp. 1309–1331, 2009.

[59] H. Yuan, "gmsh2nek. Mathematics and Computer Science Division of Argonne National Laboratory.." https://github.com/yhaomin2007/Nek5000/. (Retrieved on 04/10/2019).

[60] H. Weller, G. Tabor, H. Jasak, and C. Fureby, "A tensorial approach to computational continuum mechanics using object-oriented techniques," *Computers in Physics*, vol. 12, pp. 620–631, Nov. 1998.

[61] K. Mittal and P. Fischer, "Mesh smoothing for the spectral element method," *Journal of Scientific Computing*, vol. 78, no. 2, pp. 1152–1173, 2019.

[62] L. U. Schrader, L. Brandt, and D. S. Henningson, "Receptivity mechanisms in three-dimensional boundary-layer flows," *Journal of Fluid Mechanics*, vol. 618, pp. 209–241, 2009.

[63] D. Tempelmann, L. U. Schrader, A. Hanifi, L. Brandt, and D. Henningson, "Numerical study of boundary-layer receptivity on a swept wing," in *6th AIAA Theoretical Fluid Mechanics Conference*, 2011. doi:10.2514/6.2011-3294.

[64] N. Offermans, O. Marin, M. Schanen, J. Gong, P. Fischer, P. Schlatter, A. Obabko, A. Peplinski, M. Hutchinson, and E. Merzari, "On the strong scaling of the spectral element solver Nek5000 on petascale systems," in *Proceedings of the Exascale Applications and Software Conference 2016*, pp. 1–10, 2016.

[65] J. Ahrens, B. Geveci, and C. Law, "Paraview: An end-user tool for large data visualization," *The Visualization Handbook*, vol. 717, 2005.

[66] U. Ayachit, *The Paraview guide: A Parallel Visualization Application*. Kitware, Inc., 2015.

[67] K. Pascioni, L. N. Cattafesta, and M. M. Choudhari, "An experimental investigation of the 30P30N multi-element high-lift airfoil," in *20th AIAA/CEAS Aeroacoustics Conference*, 2014. doi:10.2514/6.2014-3062.

[68] T. M. Faure, P. Adrianos, F. Lusseyran, and L. Pastur, "Visualizations of the flow inside an open cavity at medium range Reynolds numbers," *Experiments in Fluids*, vol. 42, no. 2, pp. 169–184, 2007.

[69] T. M. Faure, L. Pastur, F. Lusseyran, Y. Fraigneau, and D. Bisch, "Three-dimensional centrifugal instabilities development inside a parallelepipedic open cavity of various shape," *Experiments in Fluids*, vol. 47, no. 3, pp. 395–410, 2009.

[70] C. Douay, L. Pastur, and F. Lusseyran, "Centrifugal instabilities in an experimental open cavity flow," *Journal of Fluid Mechanics*, vol. 788, pp. 670–694, 2016.

[71] D. Barkley, M. G. M. Gomes, and R. D. Henderson, "Three-dimensional instability in flow over a backward-facing step," *Journal of Fluid Mechanics*, vol. 473, pp. 167–190, 2002.

[72] J. F. Beaudoin, O. Cadot, J. L. Aider, and J. E. Wesfreid, "Three-dimensional stationary flow over a backward-facing step," *European Journal of Mechanics-B/Fluids*, vol. 23, no. 1, pp. 147–155, 2004.

[73] C. Williamson, "The existence of two stages in the transition to three-dimensionality of a cylinder wake," *The Physics of Fluids*, vol. 31, no. 11, pp. 3165–3168, 1988.

[74] C. Williamson, "Three-dimensional wake transition," *Journal of Fluid Mechanics*, vol. 328, pp. 345–407, 1996.

[75] G. Agbaglah and C. Mavriplis, "Computational analysis of physical mechanisms at the onset of three-dimensionality in the wake of a square cylinder," *Journal of Fluid Mechanics*, vol. 833, pp. 631–647, 2017.

[76] G. Agbaglah and C. Mavriplis, "Three-dimensional wakes behind cylinders of square and circular cross-section: early and long-time dynamics," *Journal of Fluid Mechanics*, vol. 870, pp. 419–432, 2019.

[77] C. H. Williamson, "Vortex dynamics in the cylinder wake," *Annual Review of Fluid Mechanics*, vol. 28, no. 1, pp. 477–539, 1996.

[78] D. Barkley and R. D. Henderson, "Three-dimensional Floquet stability analysis of the wake of a circular cylinder," *Journal of Fluid Mechanics*, vol. 322, pp. 215–241, 1996.

[79] X. Chen, J. Chen, X. Yuan, G. Tu, and Y. Zhang, "From primary instabilities to secondary instabilities in Görtler vortex flows," *Advances in Aerodynamics*, vol. 1, no. 1, p. 19, 2019.

[80] J. Ren and S. Fu, "Nonlinear development of the multiple Görtler modes in hypersonic boundary layer flows," in *43rd AIAA Fluid Dynamics Conference*, 2013. doi:10.2514/6.2013-2467.

[81] S. Winoto, D. Shah, H. Mitsudharmadi, *et al.*, "Flows over concave surfaces: Development of pre-set wavelength Görtler vortices," *International Journal of Fluid Machinery and Systems*, vol. 1, no. 1, pp. 10–23, 2008.

[82] J. Ren and S. Fu, "Secondary instabilities of Görtler vortices in high-speed boundary layer flows," *Journal of Fluid Mechanics*, vol. 781, pp. 388–421, 2015.

[83] J. Floryan, "On the Görtler instability of boundary layers," *Progress in Aerospace Sciences*, vol. 28, no. 3, pp. 235–271, 1991.

[84] W. S. Saric, "Görtler vortices," *Annual Review of Fluid Mechanics*, vol. 26, no. 1, pp. 379–409, 1994.

[85] M. Acarlar and C. Smith, "A study of hairpin vortices in a laminar boundary layer. Part 2: Hairpin vortices generated by fluid injection," *Journal of Fluid Mechanics*, vol. 175, pp. 43–83, 1987.

[86] A. Haidari and C. Smith, "The generation and regeneration of single hairpin vortices," *Journal of Fluid Mechanics*, vol. 277, pp. 135–162, 1994.

[87] M. Acarlar and C. Smith, "A study of hairpin vortices in a laminar boundary

layer. Part 1: Hairpin vortices generated by a hemisphere protuberance," *Journal of Fluid Mechanics*, vol. 175, pp. 1–41, 1987.

[88] T. Kolev, "Ceed-ms34: Improve performance and capabilities of ceed-enabled ecp applications on summit/sierra," Lawrence Livermore National Lab (LLNL) Technical Report LLNL-TR-808458, 2020.

Appendix A

Velocity Contours

A.1 Instantaneous Velocity

Figure A.1: Instantaneous X-velocity contours for flow around the 30P30N airfoil for $Re_c = 8.32 \times 10^3$ and AOA = 4° at $tU_\infty/c = 69.8$ (2D case)

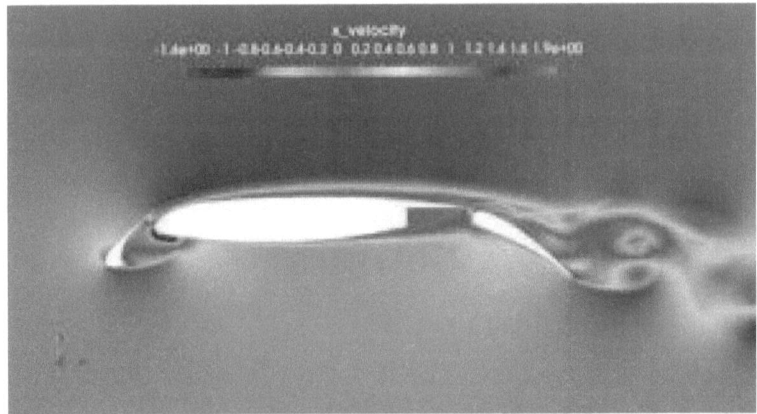

Figure A.2: Instantaneous X-velocity contours for flow around the 30P30N airfoil for $Re_c = 1.27 \times 10^4$ and AOA = 4° at $tU_\infty/c = 74.9$ (2D case)

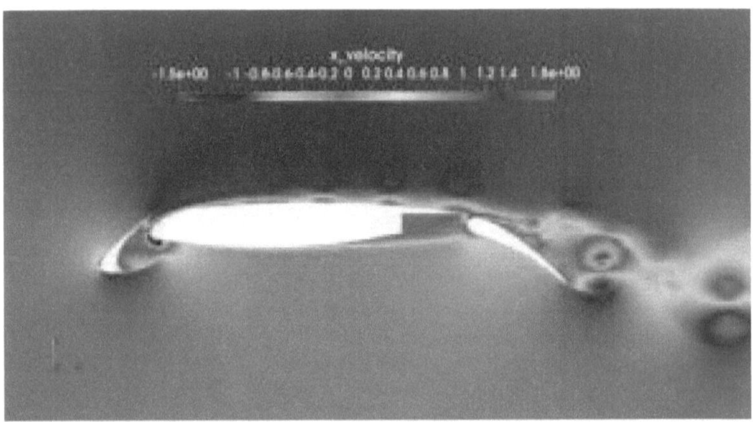

Figure A.3: Instantaneous X-velocity contours for flow around the 30P30N airfoil for $Re_c = 1.83 \times 10^4$ and AOA = 4° at $tU_\infty/c = 70.6$ (2D case)

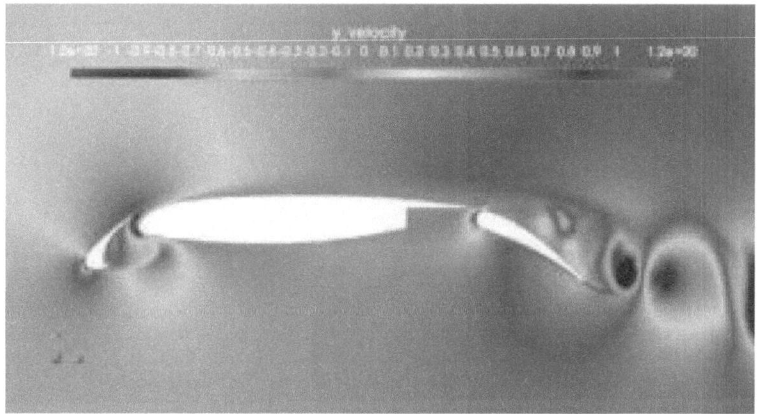

Figure A.4: Instantaneous Y-velocity contours for flow around the 30P30N airfoil for $Re_c = 8.32 \times 10^3$ and AOA $= 4°$ at $tU_\infty/c = 69.8$ (2D case)

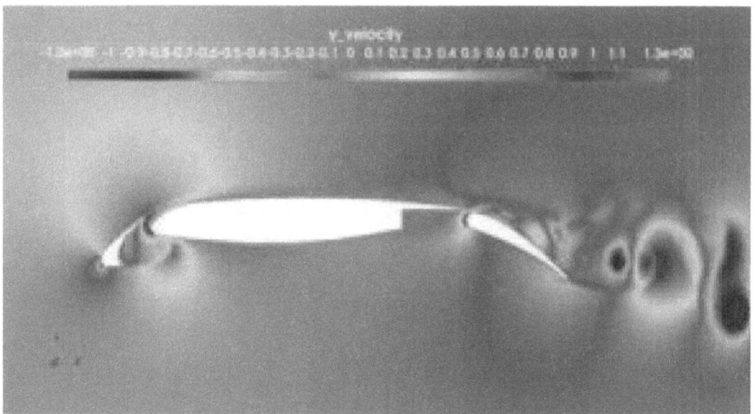

Figure A.5: Instantaneous Y-velocity contours for flow around the 30P30N airfoil for $Re_c = 1.27 \times 10^4$ and AOA $= 4°$ at $tU_\infty/c = 74.9$ (2D case)

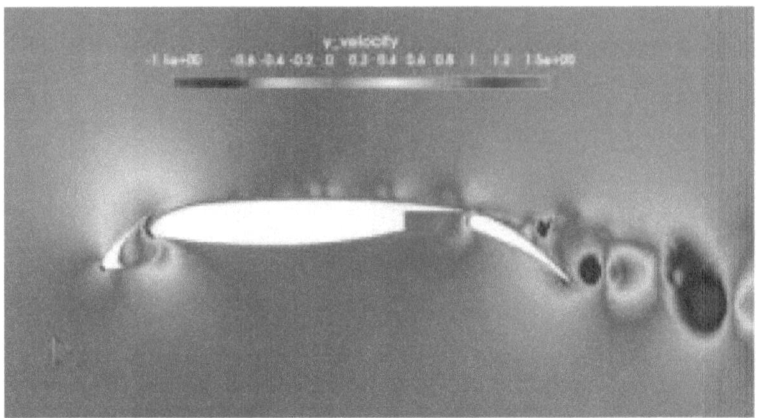

Figure A.6: Instantaneous Y-velocity contours for flow around the 30P30N airfoil for $Re_c = 1.83 \times 10^4$ and AOA = 4° at $tU_\infty/c = 70.6$ (2D case)

Figure A.7: Instantaneous velocity magnitude contours for flow around the 30P30N airfoil for $Re_c = 8.32 \times 10^3$ and AOA = 4° at $tU_\infty/c = 69.8$ (2D case)

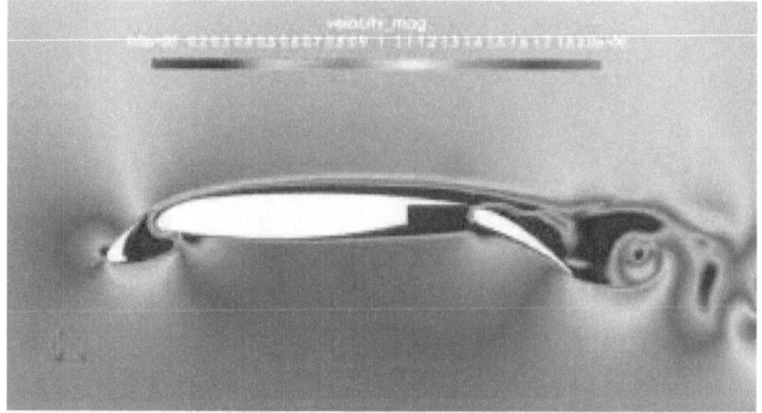

Figure A.8: Instantaneous velocity magnitude contours for flow around the 30P30N airfoil for $Re_c = 1.27 \times 10^4$ and AOA $= 4°$ at $tU_\infty/c = 74.9$ (2D case)

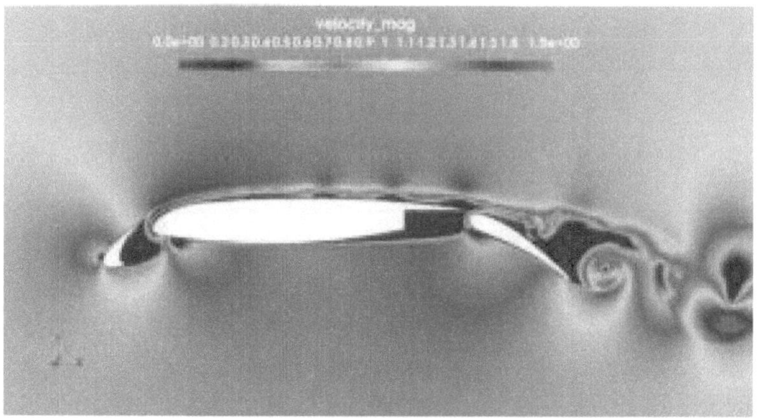

Figure A.9: Instantaneous velocity magnitude contours for flow around the 30P30N airfoil for $Re_c = 1.83 \times 10^4$ and AOA $= 4°$ at $tU_\infty/c = 70.6$ (2D case)

A.2   Time-Averaged Velocity

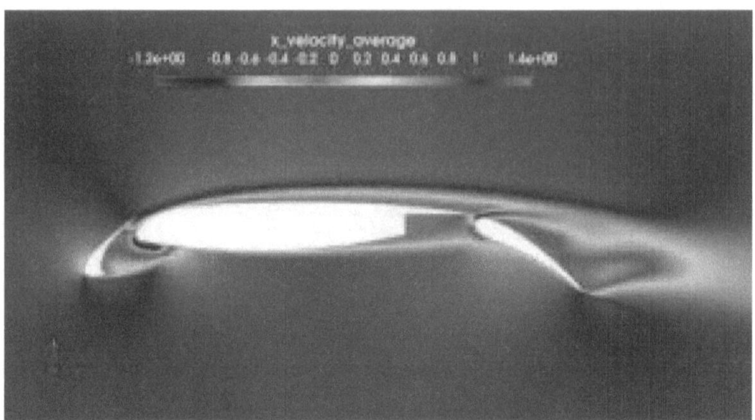

Figure A.10: Time averaged X-velocity contours for flow around the 30P30N airfoil for $Re_c = 8.32 \times 10^3$ and AOA = 4° (2D case)

Figure A.11: Time averaged X-velocity contours for flow around the 30P30N airfoil for $Re_c = 1.27 \times 10^4$ and AOA = 4° (2D case)

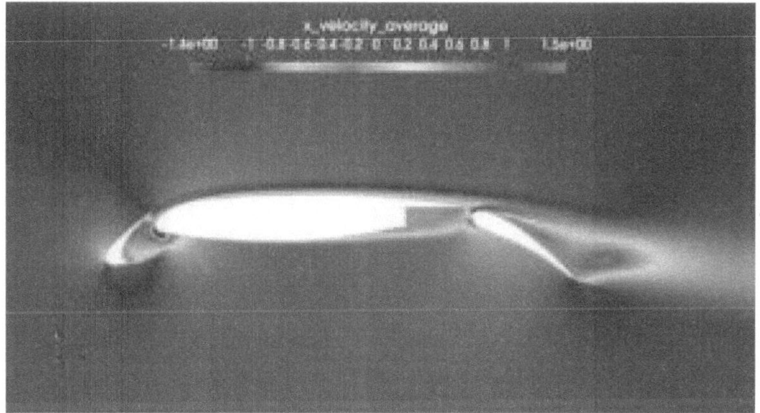

Figure A.12: Time averaged X-velocity contours for flow around the 30P30N airfoil for $Re_c = 1.83 \times 10^4$ and AOA = $4°$ (2D case)

Figure A.13: Time averaged Y-velocity contours for flow around the 30P30N airfoil for $Re_c = 8.32 \times 10^3$ and AOA = $4°$ (2D case)

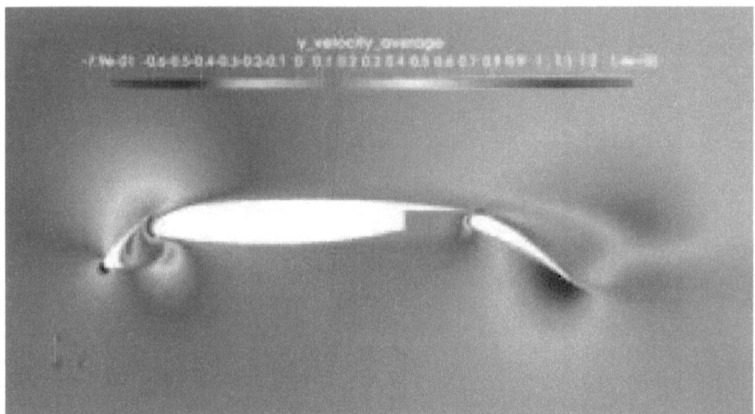

Figure A.14: Time averaged Y-velocity contours for flow around the 30P30N airfoil for $Re_c = 1.27 \times 10^4$ and AOA $= 4°$ (2D case)

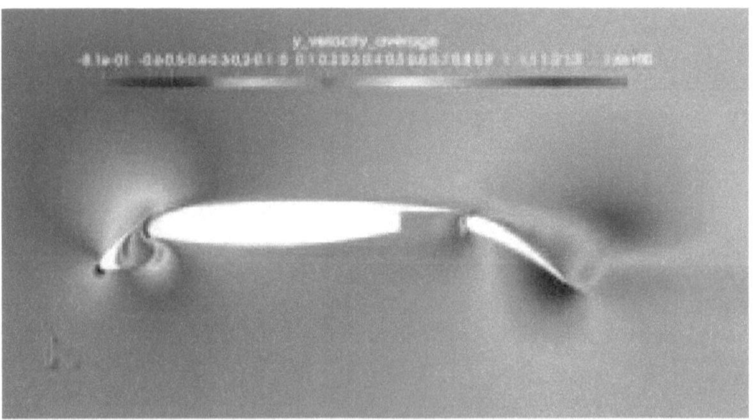

Figure A.15: Time averaged Y-velocity contours for flow around the 30P30N airfoil for $Re_c = 1.83 \times 10^4$ and AOA $= 4°$ (2D case)

Figure A.16: Time averaged velocity magnitude contours for flow around the 30P30N airfoil for $Re_c = 8.32 \times 10^3$ and AOA = 4° (2D case)

Figure A.17: Time averaged velocity magnitude contours for flow around the 30P30N airfoil for $Re_c = 1.27 \times 10^4$ and AOA = 4° (2D case)

Figure A.18: Time averaged velocity magnitude contours for flow around the 30P30N airfoil for $Re_c = 1.83 \times 10^4$ and AOA = 4° (2D case)

Appendix B

Pressure Contours

B.1 Instantaneous Pressure

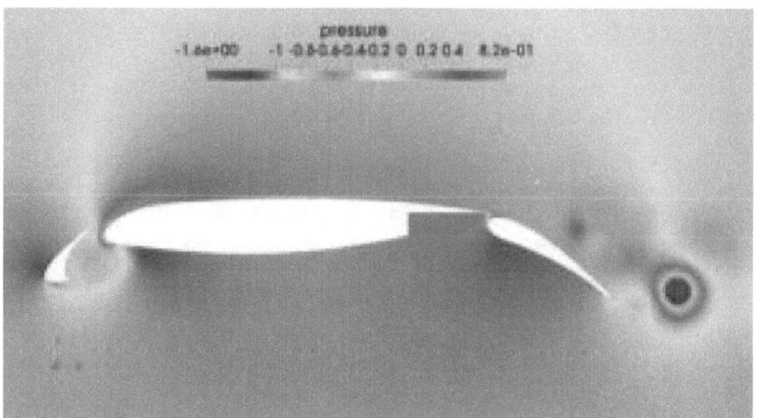

Figure B.1: Instantaneous pressure contours for flow around the 30P30N airfoil for $Re_c = 8.32 \times 10^3$ and AOA = 4° at $tU_\infty/c = 69.8$ (2D case)

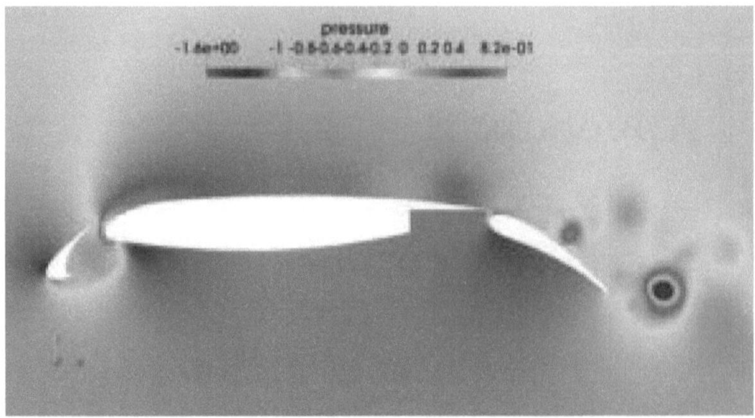

Figure B.2: Instantaneous pressure contours for flow around the 30P30N airfoil for $Re_c = 1.27 \times 10^4$ and AOA $= 4°$ at $tU_\infty/c = 74.9$ (2D case)

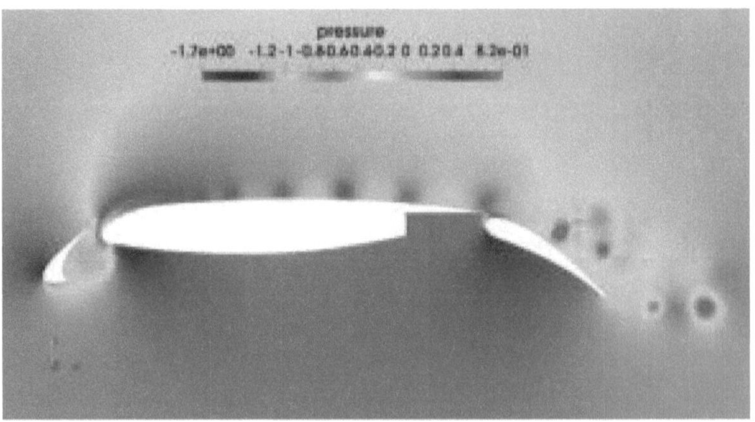

Figure B.3: Instantaneous pressure contours for flow around the 30P30N airfoil for $Re_c = 1.83 \times 10^4$ and AOA $= 4°$ at $tU_\infty/c = 70.6$ (2D case)

B.2 Time-Averaged Pressure

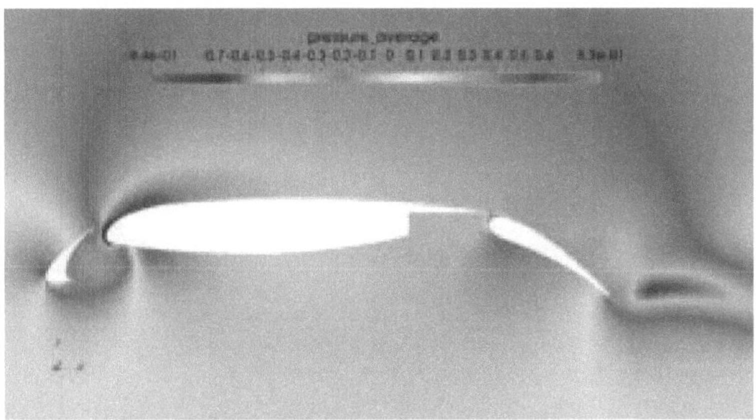

Figure B.4: Time averaged pressure contours for flow around the 30P30N airfoil for $Re_c = 8.32 \times 10^3$ and AOA = 4° (2D case)

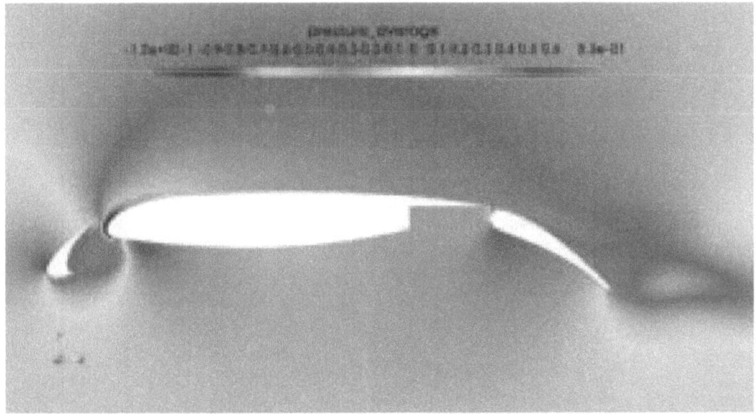

Figure B.5: Time averaged pressure contours for flow around the 30P30N airfoil for $Re_c = 1.27 \times 10^4$ and AOA = 4° (2D case)

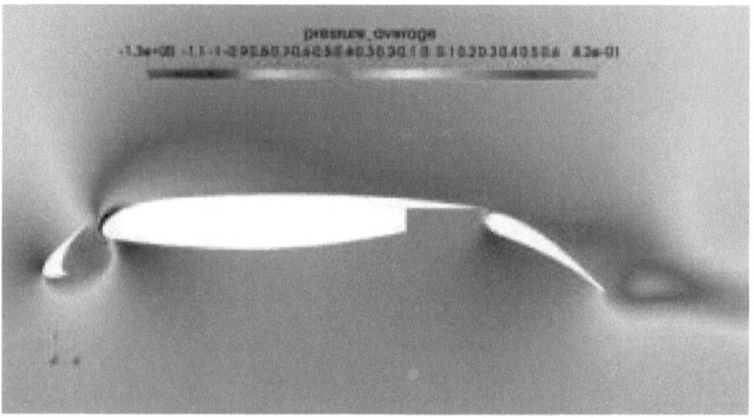

Figure B.6: Time averaged pressure contours for flow around the 30P30N airfoil for $Re_c = 1.83 \times 10^4$ and AOA = 4° (2D case)

Appendix C

Görtler Vortices

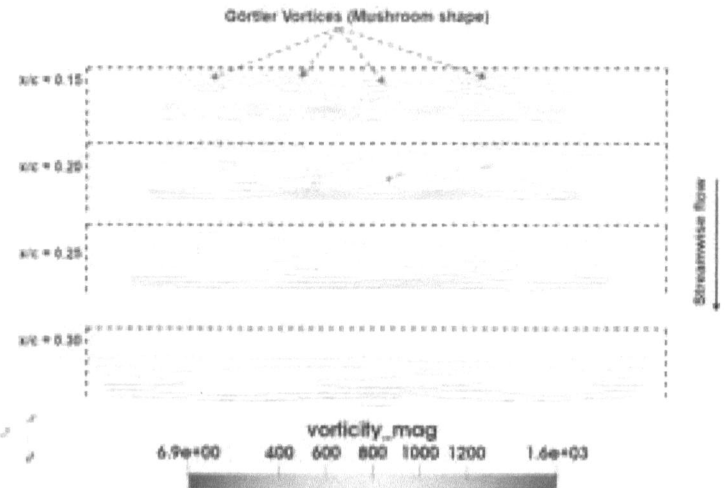

Figure C.1: Conventional "Mushroom shape" of Görtler vortices identified by cross-sectional contours of vorticity magnitude at different locations in streamwise direction for flow around the 30P30N three-element wing at $Re_c = 8.32 \times 10^3$ and $tU_\infty/c = 15.87$ (3D case)

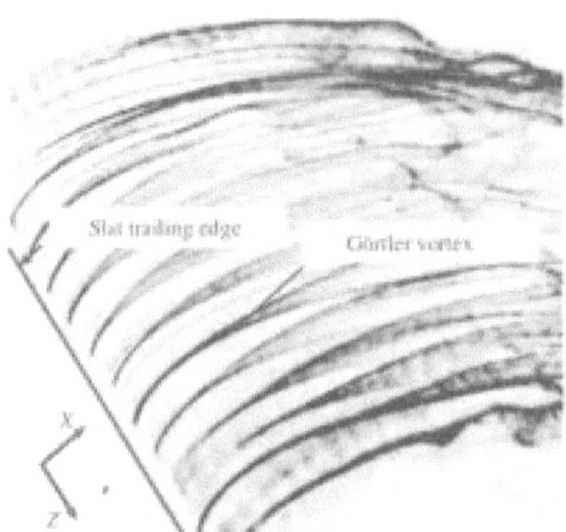

Figure C.2: Görtler vortices identified through hydrogen bubble flow visualization (isometric view) for flow around the 30P30N three-element wing at $Re_c = 8.32 \times 10^3$ [2]

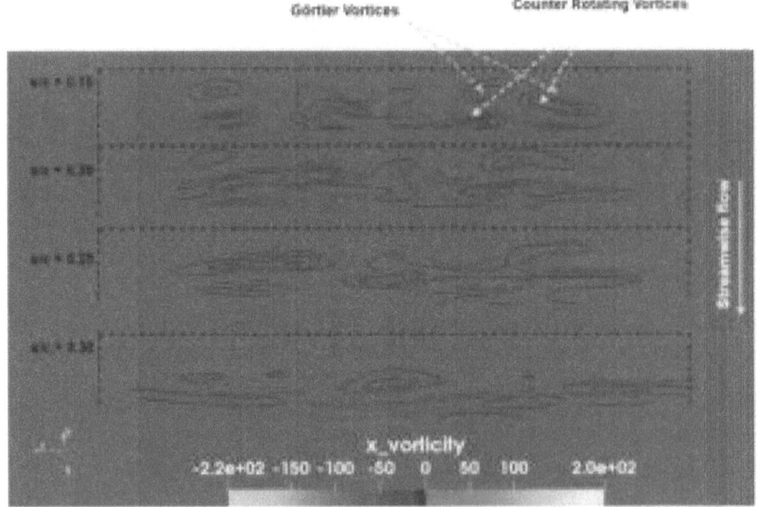

Figure C.3: Counter-rotating vortices identified beneath Görtler vortices by cross-sectional contours of X-vorticity at different locations in streamwise direction for flow around the 30P30N three-element wing at $Re_c = 8.32 \times 10^3$ and $tU_\infty/c = 15.87$ (3D case)

Figure C.4: Instantaneous Z-vorticity contours for flow around the 30P30N wing for $Re_c = 8.32 \times 10^3$ at $tU_\infty/c = 13.8$ (3D case)

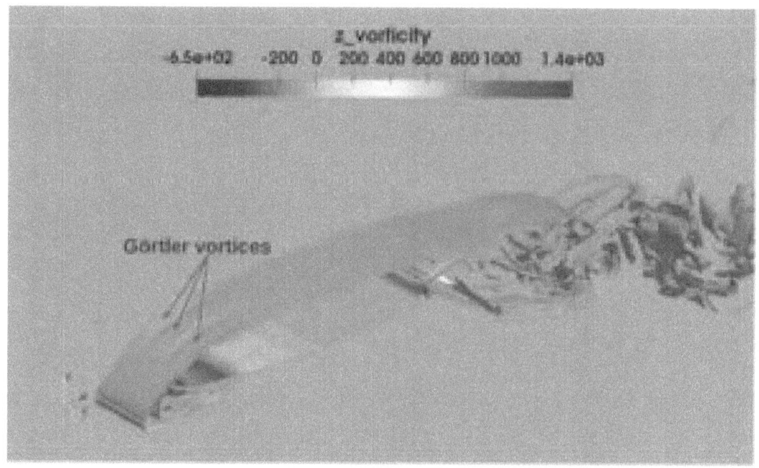

Figure C.5: Instantaneous Z-vorticity contours for flow around the 30P30N wing for $Re_c = 8.32 \times 10^3$ at $tU_\infty/c = 14.38$ (3D case)

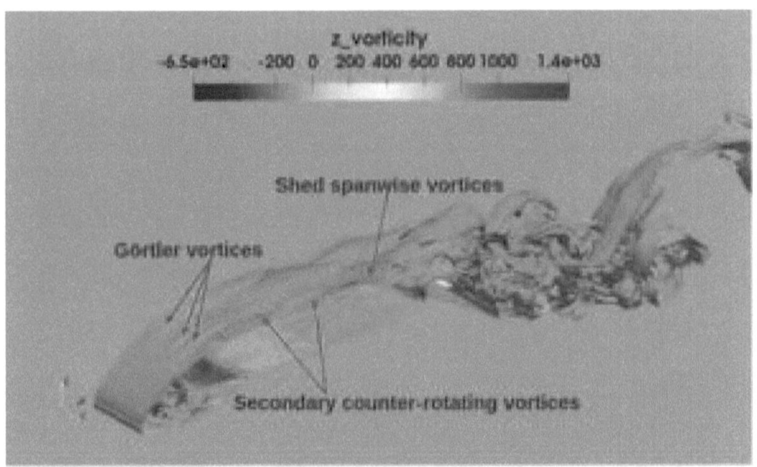

Figure C.6: Instantaneous Z-vorticity contours for flow around the 30P30N wing for $Re_c = 8.32 \times 10^3$ at $tU_\infty/c = 17.72$ (3D case)

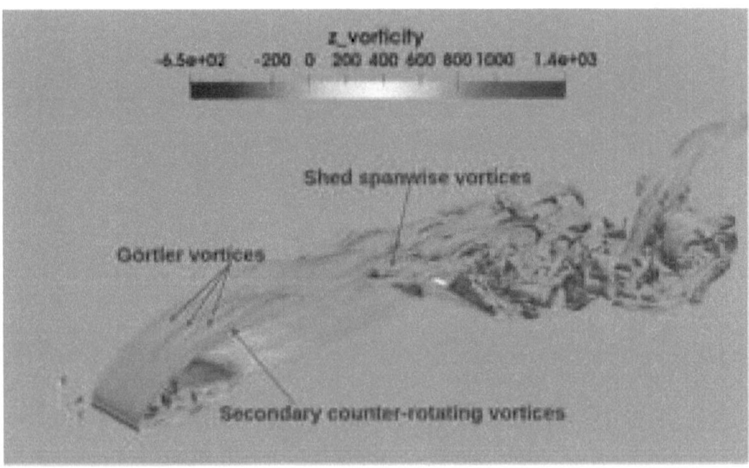

Figure C.7: Instantaneous Z-vorticity contours for flow around the 30P30N wing for $Re_c = 8.32 \times 10^3$ at $tU_\infty/c = 17.88$ (3D case)

www.ingramcontent.com/pod-product-compliance
Lightning Source LLC
LaVergne TN
LVHW041722190726
843493LV00007B/2194